Karl Mahall

Quality Assessment of Textiles

Springer-Verlag Berlin Heidelberg GmbH

Karl Mahall

Quality Assessment of Textiles

Damage Detection by Microscopy

Second Edition

With 335 Figures

 Springer

Dipl.-Ing. Karl Mahall

Heidenweg 49
40789 Monheim
Germany

ISBN 978-3-540-44072-7   ISBN 978-3-642-55645-6 (eBook)
DOI 10.1007/978-3-642-55645-6

Cataloging-in-Publication Data applied for

Bibliographic information published by Die Deutsche Bibliothek
Die Deutsche Bibliothek lists this publication in the Deutsche Nationalbibliografie; detailed bibliographic data
is available in the Internet at <http://dnb.ddb.de>.

http://www.springer.de

© Springer-Verlag Berlin Heidelberg 1993, 2003
Originally published by Springer-Verlag Berlin Heidelberg New York in 2003

Production Editor: Renate Albers, Berlin
Coverdesign: design & production, Heidelberg
Typesetting: Fotosatz-Service Köhler GmbH, Würzburg

Printed on acid-free paper   SPIN: 10889228   02/3020 ra – 5 4 3 2 1 0

Foreword

Assessing the quality of textiles using textile microscopy remains one of the important instruments for permanent process improvement in the fiber, textile and apparel industries. The degree of international interlinking in the textile producing and finishing industries and their markets demands clearly defined and reproducible methods of detecting damage or defects at all process stages.

This book – **Quality Assessment of Textiles – Damage Detection by Microscopy** – has in the meantime established itself so well as "the Mahall" in research institute laboratories investigating defects, in universities and colleges, in the training of textile chemists and technologists, and in the industry and the retail trade, that it has become necessary to bring out a new edition.

This edition has been revised and supplemented by Mr. Mahall and his successor Ms. Irmhild Goebel and her staff.

Cognis, as the successor organization continuing the textile business of the former Textile Technology department of Henkel, is pleased to make this new edition available to specialists, to students and to any other interested readers.

June 2002 *Dr. U. Kloubert* (Cognis Deutschland GmbH & Co. KG)
Prof. E. Finnimore (Fachhochschule Hof, Germany)

Foreword to the First Edition

Quality is the decisive criterion by which textile industry is measured in the international competition. Today this is particularly true.

Short fashion cycles lead to frequent article changes in production, technological progress requires continual adaptation of the production processes; high and above all constant quality of the textiles remains an indispensable requirement. Today, quality is no longer (mis)understood as the result of quality control or successful fault correction, but as the logical result of all chemical and physical or human interventions in the production process; their registration and representation in the form of quality management systems becomes more and more important. Especially in the multi-stage process of textile production and textile finishing, often carried out by several specialized companies, it is very difficult to trace back quality deficiencies in textiles – in particular hidden faults – to their true cause. However, this is the precondition to efficiently eliminate faults and to guarantee correct process control.

In his book Karl Mahall describes the damage which can occur in certain fibrous raw materials and during production and storage of textiles; for this purpose he has carefully chosen typical practical examples which he encountered in connection with textile auxiliaries during the 40 years which he has been working for the Henkel KGaA. In particular it is demonstrated how microscopic test methods can provide decisive hints at the cause of hidden faults in textiles.

For many years Karl Mahall has been recognized as an outstanding expert in the area of damage detection and as the author of a large number of lectures and publications. The typical form of representation of his considerations and examinations which invite his readers and listeners to follow his thoughts and to understand the results, has also been maintained in this book as far as the systematic order of his rich experience allows this. We therefore hope that this book will be useful as a manual and reference book for all people in industry, trade and education who have to deal with damage detection and quality assurance of textiles. However, this book should also be recommended to all other textile experts who want to get an insight into this special field; they will enjoy the short expressive statements and the impressive photographs.

Dr. G. K. Klement (Henkel KGaA)
Dr. W. Loy (Head of the Staatliches Berufsbildungszentrum
Textil-Bekleidung; Münchberg/Naila, Germany)

Preface of the Author

This book is a compendium of the experience I have gathered in the field of practical textile microscopy and as such is primarily directed at practitioners. In nearly 40 years of laboratory work, I have repeatedly applied and further developed the methods for detecting defects in textiles. A major part of the results of my work has been made available to the public in papers and published in technical journals, which are listed in the bibliography.

The first edition of my book, published in German by the technical publishing house Schiele & Schön, Berlin, in 1989, went out of print within a relatively short time, since the book became well established as a standard work in the textile industry, textile institutes and technical colleges. In 1993, Springer Verlag of Heidelberg therefore published my book in English, making it more internationally accessible. Since there is still a demand for this book, it seemed appropriate to bring out a second edition.

Mr. Hans-Josef Wieser of Cognis Deutschland GmbH & Co.KG was especially supportive in encouraging me to prepare this second edition. Given the increasing degree of globalization, Cognis prefers the English version of my book. Springer Verlag, Heidelberg, has therefore agreed with Cognis to publish the second edition of this book in English.

Besides minor improvements and corrections, the new edition contains a new chapter:
Poultry Feathers as Filling Material for Bedding and Textiles – Analysis of Faults.
The reason for its inclusion is that natural feathers and down are not only used as a filling material for bedding but also for garments, such as anoraks, coats and sleeping bags.

Some of the dyes for dyeing tests and damage reactions are no longer available or their names have changed. I am indebted to Ms. Irmhild Goebel and Mr. Berthold Popp for testing the substitute dyes on the market as to their usability.

I feel particularly obliged to Prof. Elizabeth Finnimore, a scientist for many years at the DWI (Deutsches Wollforschungsinstitut, Aachen), and now a professor of textile chemistry and textile testing at the Fachhochschule Hof, Germany; with her final terminological check, she put the finishing touch to this book.

I would also like to thank the publishing house Springer Verlag, Heidelberg, for their cooperation, in particular regarding technical wishes with respect to the printing and good reproduction of color photographs.

Monheim, June 2002   *K. Mahall*

Preface of the Author to the First Edition

This book represents a summary of my experience in the area of practical textile microscopy and is mainly destined for practical experts. In the course of almost 40 years in the laboratories of the Henkel KGaA, Düsseldorf, I have used and developed these test methods for damage detection in textiles. An essential part of the results of my work has been published in lectures and in different scientific journals. The positive feedback to these publications has again and again shown that there is a strong interest in a summarizing overview.

For the detection of textile damage it is very important to know the effects that specific damaging actions have on the individual fiber types. The structure of this monograph follows this basic thought. In order to simplify the utilization as a reference book, at the end of the book after the bibliography you will find an alphabetic register of the pictures and a detailed index. The laboratory equipment, chemicals, reagents and dyes required for microscopic damage analysis are listed in an appendix. I hope this book will allow all those who want to work in this area to find an easy access.

After many years of practical work – in spite of all my experience with lectures and publications – writing a book was a special challenge. Dr. Klement from the Henkel KGaA, Düsseldorf, has decisively inspired me to write this monograph. Without him this book would certainly never have been accomplished. Further, I would like to thank Dr. Veitenhansl and Dr. Schlüter for their critical support in the elaboration of the manuscript and Mr. K. Siekmann for his valuable editorial work.

I also wish to thank the translators of the German edition of this book, Mrs. Jutta Müller and Mrs. Heike Kähler (Henkel KGaA, Sprachendienst), for their work, which was terminologically supported by Dr. Ulrich Rall, former head of the department Anwendungstechnik/Textilhilfsmittel at Henkel KGaA.

I feel particularly obliged to Prof. Dr. Elizabeth Finnimore, a scientist for many years at the DWI (Deutsches Wollforschungsinstitut, Aachen), and now a professor of textile chemistry at the Fachhochschule Coburg/Münchberg, Germany; with her final terminological check, she put the finishing touch to this book.

I would also like to thank the publishing house Springer Verlag, Heidelberg, for good cooperation, in particular regarding technical wishes with respect to printing and good reproduction of color photographs.

I am grateful to the Henkel KGaA for financial support during my preparative work for this book.

Monheim, October 1993   *K. Mahall*

Contents

1 Fundamentals and Priming

Inappropriate treatment of textiles during production and use can cause chemical, mechanical and thermal damage or damage due to microorganisms; each has different effects and greatly reduces the serviceability of the textiles. Usually the cause of damage cannot be determined by purely visual evaluation. Textile microscopy, in contrast, often provides exact results within a very short time, which can be decisive in the correction of future production.

Practical textile microscopy can be used in order to clarify numerous difficulties which occur during the production and finishing of textiles [1]. The identification of chemical fiber materials becomes more and more complicated due to the increasing variety of new fibers and their use for the production of widely varying fiber blends. Today it is no longer possible to distinguish different man-made fibers, especially in the case of dyed articles [2], without microscopic examination in combination with dissolving, swelling and staining reactions. When dealing with complaints of damage to textiles, extensive practical experience in all areas of textile production is required in order to evaluate the results of the microscopic examination in such a way that they are useful for practical application.

1.1 Necessary Equipment

For practical textile microscopy, simple working and/or standard microscopes with polarization equipment are sufficient, especially since this equipment must often be operated not only by trained technical staff but also by unskilled workers. An optical device with approx. 100- to 800-fold magnification is sufficient. However, the resolution power of the microscope is more important than the magnification. This is the capacity of a lens system to produce a distinct image of two close object points. Thus the resolution power depends on the quality of the lenses which produce a magnified, reversed image of the sample. With an ocular, this image becomes visible to the eye, but due to the ocular magnification there is no further detail resolution. The basic equipment should also include a camera as well as a variable temperature microscope. The latter is necessary in order to be able to specify the melting range of synthetic fibers. The equipment and reagents are listed in the appendix.

1.2 Preliminary Examination of Textile Test Material

Preparation for microscopic examination includes a detailed fabric inspection which determines whether damage is sporadic or occurs regularly throughout the textile fabric and whether individual pieces or the entire lot are affected. Furthermore, it is important to determine whether the damage can be recognized best in transmitted, reflected, diagonal or UV light. A preliminary examination should determine, for example, whether streaks run parallel to the thread, diagonal or maybe even in a bow to the weft or warp direction and also whether the damage affects only the weft yarns or only the warp yarns. In addition to microscopic examination described below, these observations are of great importance to determine the cause of damage.

The next step is the examination with a magnifying glass and/or a low-magnification microscope or stereo microscope. With a magnifying glass the object is seen in reflected light, while a stereo microscope with reflected and transmitted light equipment with 6- to 40-fold or 12- to 80-fold magnification permits examination under transmitted, reflected, diagonal or combined light. Since with a stereo microscope the working distance is large for all magnifications, it is indispensable for most preparation work.

Preliminary examination also includes large imprints, see chapter 1.6. Only after all possible preliminary examination has been carried out on the textile fabric, should splitting into yarns or individual fibers begin. In this case, the fiber composition should first be checked under the light microscope before the damaged area is examined.

1.3 Classical Examination Methods

In practical textile microscopy, the fibers are usually immersed in optically neutral liquids such as glycerol/water, paraffin oil etc. and examined under the light microscope in transmitted light. Since textiles are more or less translucent, the lower fiber half shimmers through and often complicates surface examination. This applies especially to fibers with a distinct surface structure, e.g. wool and hair, and in particular if they have an additional internal structure, e.g. a medulla. The medulla looks black in transmitted light because of the pigments as well as the air included in and between the cells. In the case of dark dyed fibers, e.g. navy blue and black, the surface is rather unclear.

Fig. 1
Page 4

In 1927 Herzog suggested that imprints of animal fibers and hairs in Krönig cover glass mounting cement (mixture of rosin and wax) [3] should be produced, in order to allow a clearer recognition of the arrangement and shape of the epidermal cells. In this microscopic examination method, cover glass mounting cement is liquefied by heating and then spread in a thin layer on a preheated microscope slide. A uniform layer thickness is obtained by reheating the microscope slide. The fibers are immersed in the soft or liquid cover glass mounting cement and, after cooling, drawn off by hand or by means of tweezers. The imprint is then examined under the microscope. Cover glass mounting cement is relatively hard so that there is no danger of subsequent deformation of the imprints, Fig. 1.

Figure 2 shows a hair, which was immersed in an optically neutral liquid, in transmitted light. The surface structure cannot be recognized, but the internal structure, i.e. the medulla, is visible. Figure 3 shows a negative imprint of the same hair in cover glass mounting cement. The epidermal cells can be clearly recognized on this imprint. This example demonstrates the high performance of negative imprint techniques for examination of objects under the microscope. Fig. 2, 3 Page 4

However, there are some disadvantages associated with imprints of individual fibers in Krönig cover glass mounting cement. Due to the hardness of the mounting cement the fibers break when they are drawn off, i.e. the outer regions are indistinct. In addition, cover glass mounting cement is only suitable for individual fibers, especially for wool and hair, but not for larger textile fabrics since it is almost impossible to produce mounting cement films with a uniform thickness.

1.4 Preparation of Negative Imprints with the Size of Microscope Slides

Reflected light microscopes are suitable for surfaces which are plane and opaque but reflect light (e.g. metal surfaces). The surfaces of textile fabrics are more or less hairy, especially if staple fiber yarns were used for their preparation.

Therefore, demands with respect to depth of focus of the reflected light microscopes are particularly high. Since, in addition, the textiles are more or less translucent, light reflection is insufficient. Thus reflected light microscopes do not always produce a satisfying image of the surface of textile fabrics, especially in the case of hairy material. In such cases imprints of these surfaces, which are produced with the aid of different ductile masses, e.g. cover glass

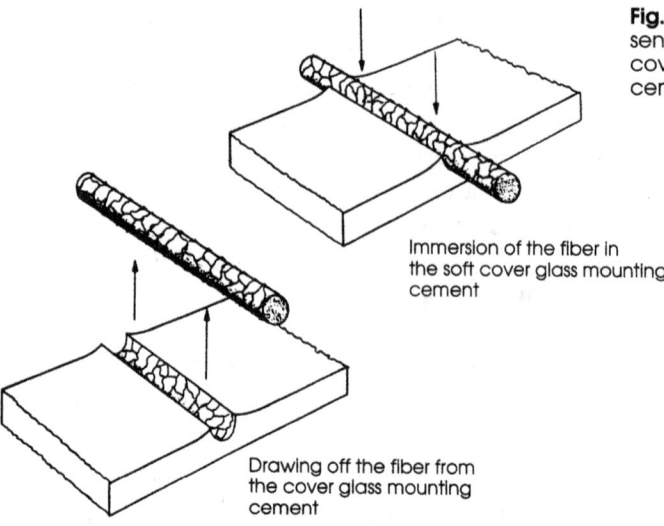

Fig. 1. Schematical representation of a fiber imprint in cover glass mounting cement.

Immersion of the fiber in the soft cover glass mounting cement

Drawing off the fiber from the cover glass mounting cement

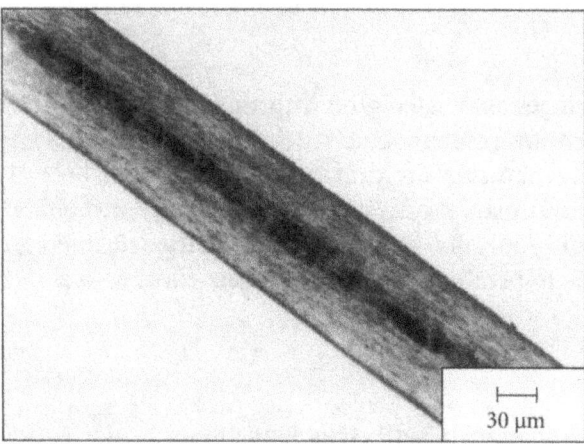

30 µm

Fig. 2. Hair, immersed in glycerol/water, in transmitted light. The surface structure cannot be recognized, but the internal structure is visible (medulla).
Unless stated otherwise, all subsequent fiber preparations were produced according to this method.

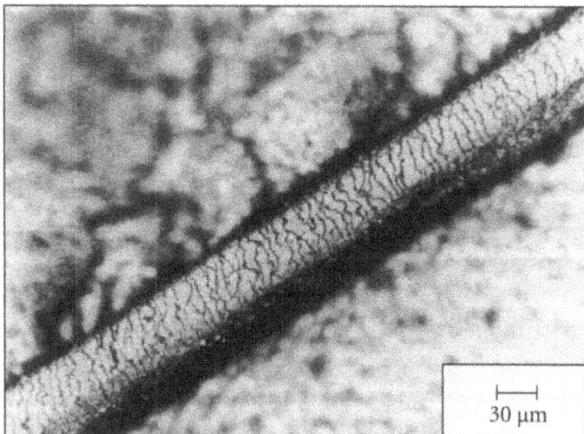

30 µm

Fig. 3. The same hair as in Fig. 2, but as an imprint in cover glass mounting cement. The surface structure is clearly visible, but it is almost impossible to recognize the internal structure of the hair.

mounting cement, glue, gelatin, polystyrene or polypropylene [4–7], can be used.

Thermoplastic films made from polystyrene and polypropylene are suitable for all samples which are resistant up to a temperature between 105 and 120 °C [8]. They are preferred to other immersing media because from the outset they form films with a uniform thickness. These imprints can be produced in a relatively simple and fast way with the size of microscope slides for microscopic examination in transmitted light.

Examination of negative imprints has the following advantages when compared to the examination of original objects:

– No reflected light microscope is needed. The surface imprints on thermoplastic films can be examined in transmitted light similarly to the usual fiber samples.

– The depth of focus is much better.

– The imprints can be magnified under the microscope up to the optimal resolution.

– The surface structure of the fibers and/or the damage to individual fibers can be clearly seen, even on the imprint of dark-coloured material, since dyes have no influence in this case.

– The luster of the material, which can often be disturbing during the examination of defects, has no influence.

– Additional fiber preparations are often unnecessary, i.e. the object remains undamaged.

Materials and equipment for the preparation of imprints on thermoplastic films with the size of microscope slides [9] are listed in the appendix.

For examination, textile fabrics are spread on the film and pressed between two polished metal plates by means of screw clamps. The plates are then placed in a drying cabinet at a temperature of 105 °C (if necessary up to a maximum of 120 °C). After 20–30 minutes the plates are removed from the drying cabinet and cooled as fast as possible (open window, refrigerator). Contours vanish when cooling takes place too slowly. After cooling, the textile material is drawn off by hand or by means of tweezers. The imprint is then examined under the microscope in transmitted light, Fig. 4 and 5. It can sometimes be advantageous to use diagonal light because it helps to produce plastic images. This procedure provides clear, distinct images of the surface structure of the textile

Fig. 4–13
Page 7–10

fabrics, Fig. 6–13, which can be magnified as desired under the light microscope up to the optimal resolution, Fig. 13. In this context one disadvantage of the imprint technique has to be mentioned: the internal structure of the fibers cannot be recognized. Fig. 13, for example, shows the imprint of a dull-spun polyester fiber whose dulling pigments in the interior of the fiber cannot be recognized from the surface imprint.

1.5 Preparation of Negative Imprints of Fibers and Yarns on Thermoplastic Films

Fig. 14–20
Page 11–13

The same technique can be used to produce surface imprints of loose fiber material and of yarns on thermoplastic films [10]. The fibers or the small yarn pieces are spread on the films and then treated in the same way as described for textile fabrics in chapter 1.4. For loose fiber material and very fine yarn, thin polypropylene films (30–35 μm) have proven to be most suitable. This method avoids the danger of the sample being pressed too deeply into the film. The importance of the depth of penetration during the production of imprints is shown in Fig. 14. Figures 15–17 show how the depth of focus for individual fibers is improved by means of imprints and illustrate the advantage that with this method the color is optically neutralized. The structure of dark dyed individual fibers can only be recognized in detail by means of this technique. Figures 18–20 display imprints of yarns. It can be seen that in the imprints the thread is flattened and mainly spread on one level. In this way a very clear image of the yarn surface is obtained.

Fig. 21
Page 13

Compared to use of two microscope slides compressed by means of hose clips, Fig. 21, imprints between two polished metal plates with screw clamps have the advantage that the pressure and thus the depth of penetration of the material can be more easily varied and optimized.

1.6 Imprints of Larger Areas

Large imprints on polystyrene films provide a good insight into the surface properties of a fabric, i.e. into the structure of the textile fabric. The dye of the material is not reproduced on the imprint. Thus, via the replication method, dyeing unlevelness can be distinguished from structure-related defects and also from deposits because the latter can diffuse into the film or adhere to it. Since much textile damage results from changes in the surface, the examination of surface imprints is of special importance.

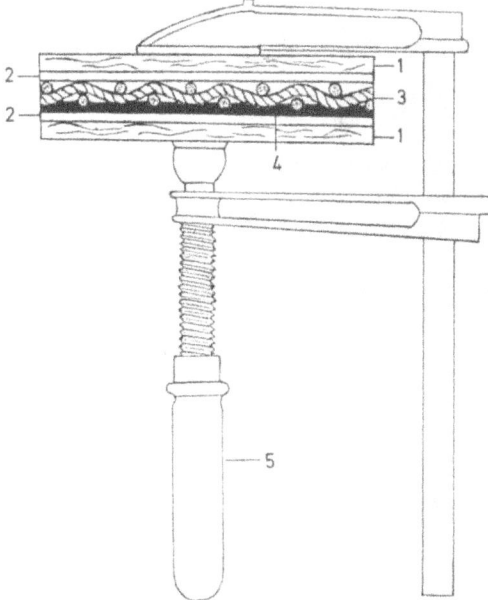

Fig. 4. Imprint of textiles on thermoplastic films compressed between polished metal plates with the size of microscope slides.
1 Wooden plate
2 Polished metal plate
3 Fabric
4 Film
5 Screw clamp

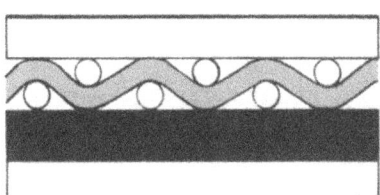

Preparation

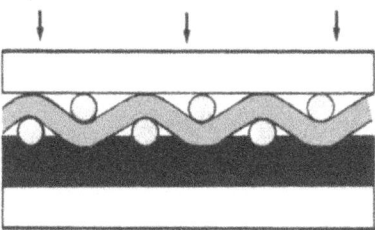

Application of pressure an heat

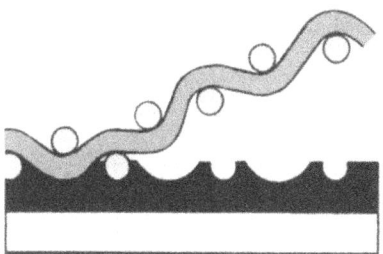

Finished imprint

Fig. 5. Preparation of a surface imprint on a thermoplastic film (schematical). Samples produced in such a way are referred to as film imprints.

Fig 6. Woollen cloth.
Left: In transmitted light;
right: In reflected light.

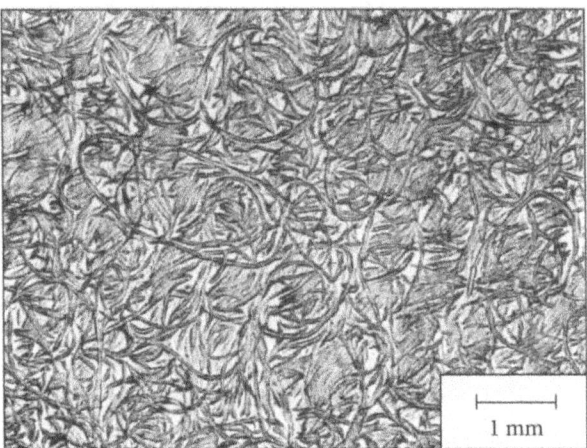

Fig. 7. Film imprint of woollen
cloth in Fig. 6.

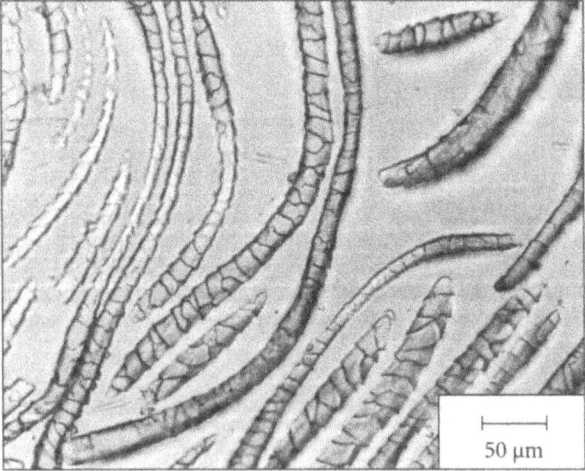

Fig. 8. Strongly magnified
section from Fig. 7.

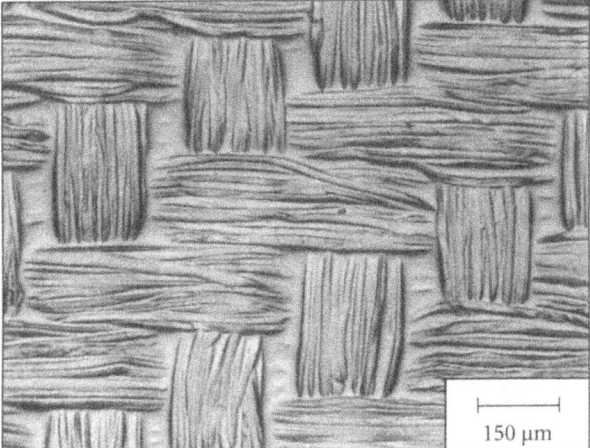

Fig. 9. Film imprint of a fabric made of pure silk.

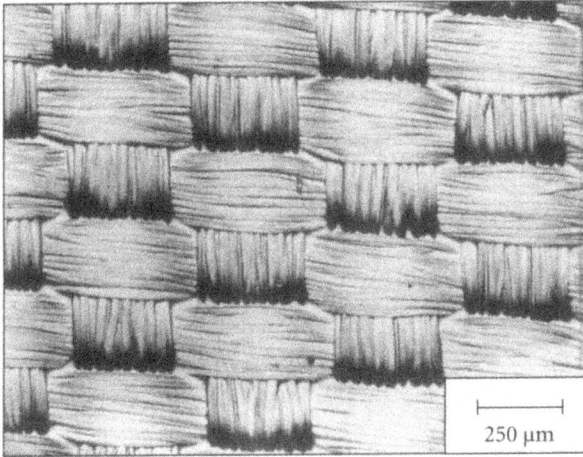

Fig. 10. Film imprint of a fabric made of viscose.

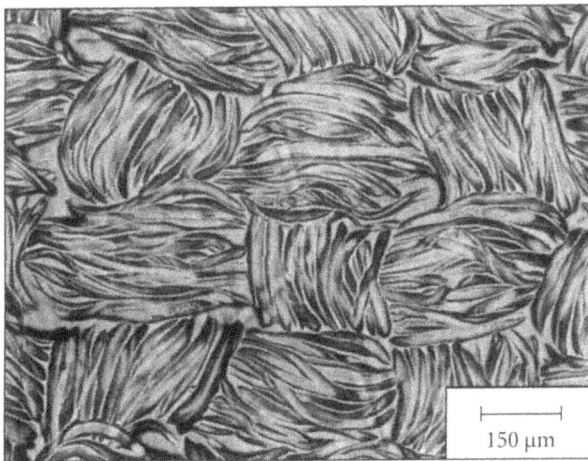

Fig. 11. Film imprint of a fabric made of textured polyester.

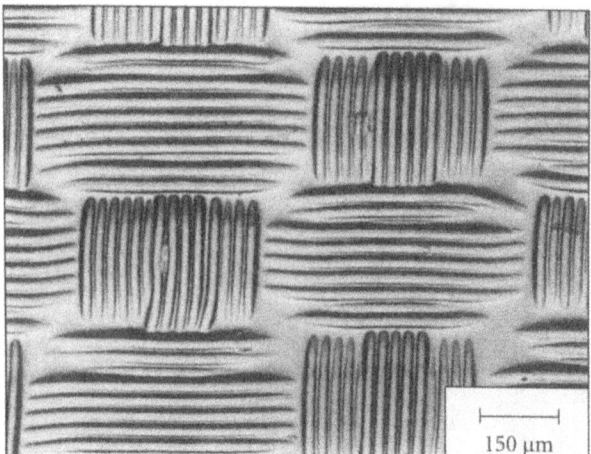

150 µm

Fig. 12. Film imprint of a fabric made of polyamide.

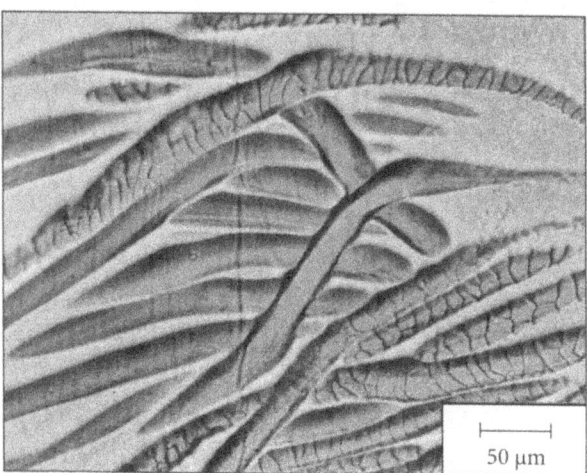

50 µm

Fig. 13. Film imprint of a fabric made of polyester/wool, strongly magnified.

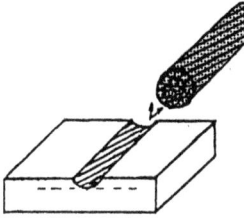

Fig. 14. The importance of the depth of penetration in a film imprint.

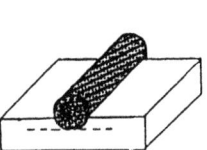

Yarn sample is not pressed deeply enough into the film; insufficient representation of the surface.

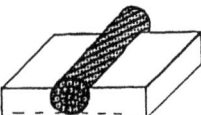

Yarn sample is pressed too deeply into the film; yarn cannot be drawn off or edges break.

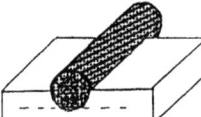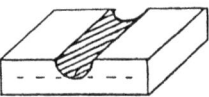

Optimal depth of penetration; yarn sample can be drawn off easily.

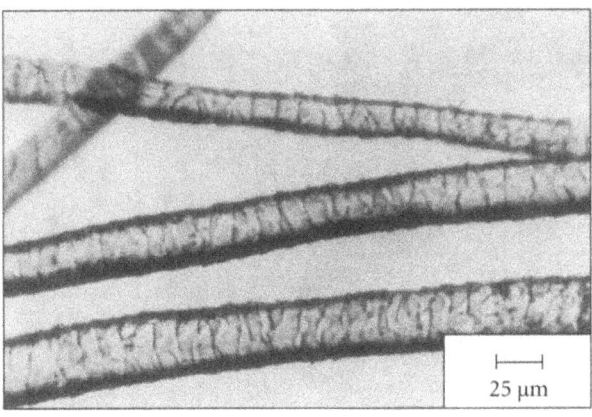

25 µm

Fig. 15. Wool fiber, embedded in glycerol/water, in transmitted light.

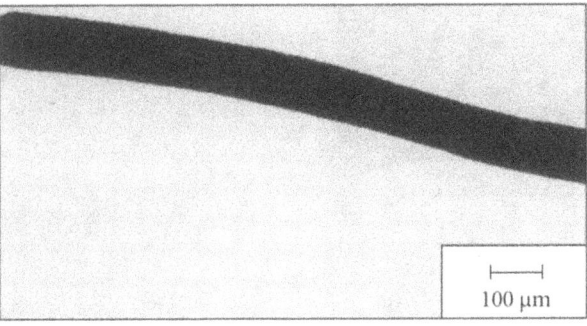

100 µm

Fig. 16. Coarse wool fiber dyed black; glycerol/water preparation, in transmitted light. The scale structure is hardly detectable.

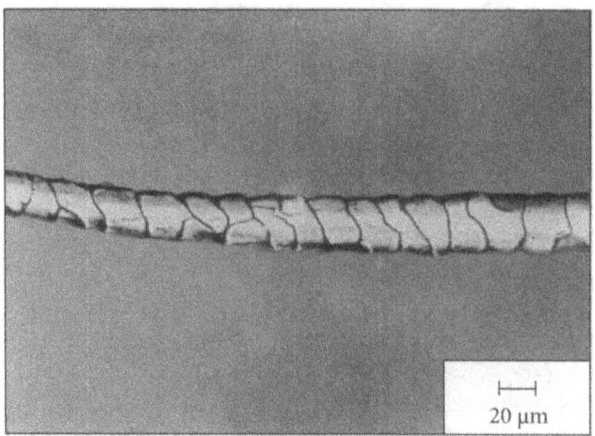

Fig. 17. Film imprint of a wool fiber; the dye is eliminated, while the scale structure can be clearly recognized.

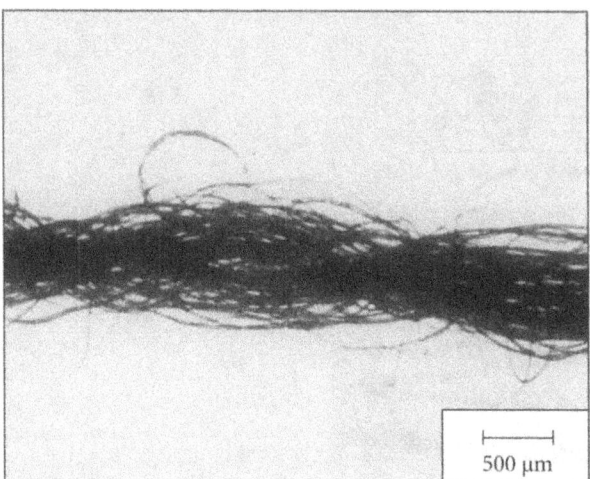

Fig. 18. Twoply wool yarn in reflected light.

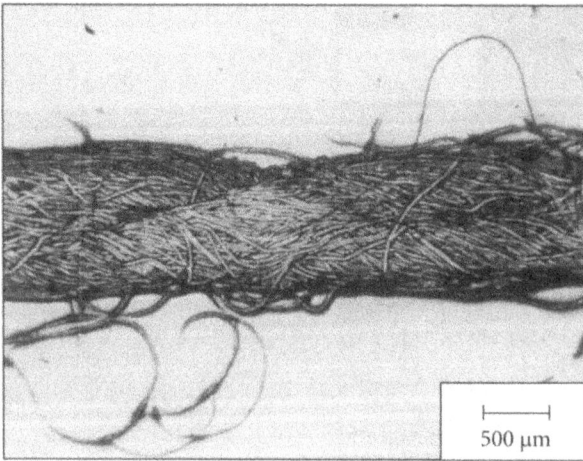

Fig. 19. The same yarn as in Fig. 18, but as a film imprint.

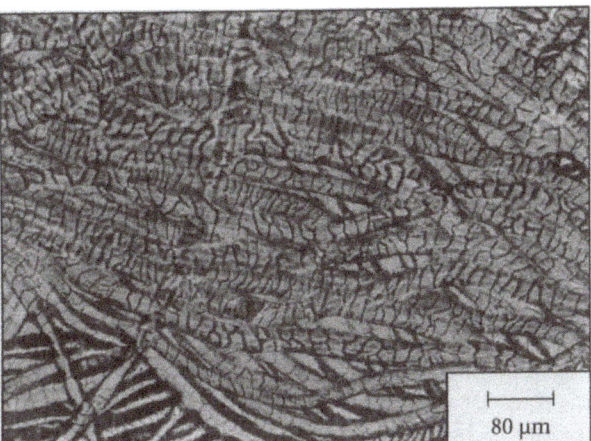

Fig. 20. Magnified section of the film imprint from Fig. 19.

Fig. 21. Preparation of a film imprint between two microscope slides.

Fig. 22. Streak analyzer for the preparation of large film imprints.

Fig. 22
Page 13
For large imprints, Atlas Devices Company (distribution in the Federal Republic of Germany by Atlas SFTS B.V., Mülheim/Ruhr) developed an apparatus for the production of large imprints in DIN A 4 size on thermoplastic films, Fig. 22.

Heating cartridges are accommodated in the lower steel plate of the so-called streak analyzer. A thin, polished stainless steel plate is attached to this part of the apparatus. It supports a 100 um thick polystyrene film with the textile fabric. This is covered with a rubber mat for sealing, and the upper steel plate is screwed on. The temperature is controlled by a thermostat (accuracy ± 0.75 °C). Using compressed air the textile fabric is pressed for 2–3 minutes into the soft polystyrene film at a temperature of 120–121 °C. Experience has shown that the pressure should lie between $1.0 \cdot 10^5 - 1.8 \cdot 10^5$ Pa (1.0 to 1.8 bar), depending on how deep the material is pressed into the film; this ultimately depends on the thickness of the textile fabric. The pressure can be varied with the aid of a pressure gauge. The water connection is located at the rear of the apparatus. The steel plates can be cooled with water so that the contours do not disappear due to slow cooling. After cooling, the textile fabric is drawn off the film which is then placed onto a black plate and examined. All large imprints described in this book were produced according to this method. Its application is explained in more detail by means of some practical examples.

1.6.1 Detection of Dyeing Unlevelness with the Aid of the Replication Method – Practical Example

Fig. 23
Page 17
A yarn-dyed fabric was used whose warp and weft consisted of polyester/wool/mohair. After weaving, a streakiness and/or barré was observed in the weft, Fig. 23.

Preliminary examination revealed that it was a uniform, clean fiber material without any incrustations or deposits. There were no signs of damage to the fiber material. Furthermore, no differences could be found in the residual grease content within the streaky parts (see chapter 1.6.2).

The large imprint on a polystyrene film clearly showed that the streakiness was not structure-related since it did not appear on the imprint. Consequently, this material defect was undoubtedly caused by dyeing unlevelness.

1.6.2 Detection of Oil and/or Grease Soiling on Textile Fabrics

During production of imprints, oily and/or grease-like substances diffuse into the softened thermoplastic films thus causing turbidity, Fig. 24. If there are differences in the spin-finish pick-up as well as in the grease, wax, paraffin or silicone content within the textile fabric, this is clearly discernible on the imprint. The detection of further deposits on textiles is described in chapter 7.

Fig. 24
Page 17

1.6.3 Detection of Structural Defects – Practical Example

Pieces of knitted fabric made of pure wool, produced from a red-dyed yarn, showed an irregular streak formation which, depending on the incident light, could be more or less clearly observed. On the imprint the streak formation was represented with the same strength and form as on the original fabric. No change occurred after extraction of the fabric with petroleum ether. For this reason, a different oil and/or paraffin add-on combined with dyeing unlevelness could not account for this problem. It could only be attributed to a structural defect, i.e. to yarn differences which cause differences in the stitch structure, Fig. 25. These lead to optical effects; in areas where there are larger gaps between the stitches, light shines through during examination with reflected light, thus giving the impression of darker dyeing. In areas where the stitches are more compact the incident light is reflected, thus giving the impression of lighter dyeing.

Fig. 25
Page 17

1.6.4 Summarizing Evaluation of Large Imprints

These practical examples demonstrate that it is often reasonable to produce large imprints in addition to the preliminary tests (chapter 1.2); they reveal whether damage must be attributed to dyeing unlevelness, structural defects or deposits.

Table 1 (see p. 16) illustrates the possible results obtainable from large imprints.

1.7 Microtome Sections

The microscopic examination of fiber, yarn or fabric cross-sections is not only used for the general evaluation of the cross-section form; it is also important for the identification of the fibers as well as for damage analysis. For the preparation of microtome sections, the fibers have to be embedded. In practical textile microscopy, use of polyester casting resin such as Leguval (Bayer) proved to be a fast embedding method. Fiber, yarn and fabric cross-sections in this book were produced with this polyester casting resin. The following formulation proved to be especially suitable for block embedding:

> 10 g Leguval E 81
> 0.1–0.2 g Leguval K 70 (harder than Leguval E 81),
> 0.3–0.5 g Benzoyl peroxide, 50%, (Akzo Chemie GmbH, Emmerich).

The material to be sectioned is bonded onto a cardboard frame with Pritt stick glue (Henkel). The small frame with fabric fragments or fiber bundles is placed diagonally into a small cardboard box with a side length of approx. 2 cm; the liquid resin is then added. The box containing the resin mixture is put into a cold drying cabinet and heated up to 80 °C. At this temperature the resin polymerizes after a period of 40 to 60 minutes. After cooling, the block is cut to shape, keeping the cut surface as small as possible. Afterwards, the small block is glued to a plastic mount with Pattex Stabilit express (Henkel). On a sliding microtome, sections of approx. 10 μm can then be produced. The sections are immersed in cedar oil (Merck) on a microscope slide, which optically eliminates the synthetic resin. Only the fiber material can now be seen under the microscope.

Table 1: Survey of the results obtained from large imprints

Delivered fabric	Fabric after extraction with petroleum ethe	Cause of damage
–	–	Dyeing defect
+	–	Oil, grease, wax or paraffin
+	+	Structural defect, possibly pigment soiling

+ Defects which can be recognized on the negative imprint;
– Defects which cannot be recognized on the negative imprint

Fig. 23. Above: Fabric with weft streakiness (polyester/wool/mohair).
Below: The same fabric, imprinted on a polystyrene film; no visible streak formation because unlevelness results from dyeing.

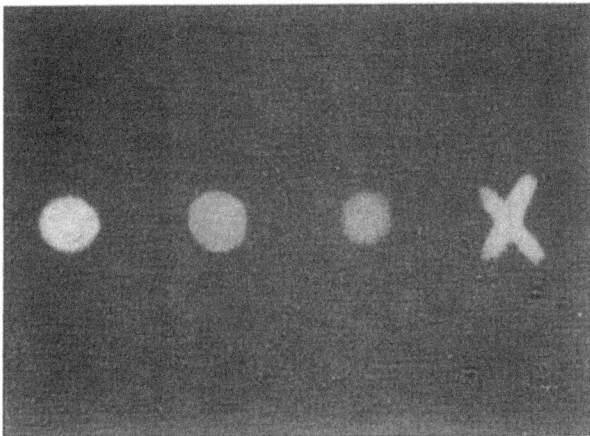

Fig. 24. Model test: Film imprint of a fabric with different stains; from left to right: Coning oil, silicone, softener, paraffin.

Fig. 25. Film imprint of a knitwear fabric of pure wool yarn. The streak formation can be attributed to yarn differences, i. e. to structural defects.

2 Chemical Damage

2.1. Chemical Damage to Wool

Fig. 26-28
Page 20-21 Wool fibers have an extremely complicated structure and characteristic scales, Fig. 26, by which they can be distinguished from other fibers. Apart from this, structural abnormalities, i.e. damage, can be detected quite easily under the microscope. Staining reactions with Cotton Blue-lactophenol [11], Neocarmin W (Fesago, Heidelberg) or Pauly reagent [12] simplify the recognition of wool damage. With Cotton Blue-lactophenol, damaged wool fibers are dyed blue, Fig. 27. With Neocarmin W – a dye solution used to distinguish various fibrous materials – undamaged wool is dyed yellow and damaged wool orange. It is very easy to perform this staining reaction. The sample is placed into the dye solution for 5 minutes at room temperature and then rinsed. However, slight damage is either not indicated or only indistinctly. The reagent responds best to badly alkaline-damaged wool, Fig. 28.

Fig. 29
Page 21 Depending on the respective degree of damage the Pauly reagent dyes damaged wool yellow, orange or reddish brown. It is unimportant whether the damage is chemical, biological or mechanical, Fig. 29. Compared to other staining reagents, the Pauly reagent is advantageous because it facilitates differentiation according to the degree of damage.

2.1.1 The Pauly Reaction

During the Pauly reaction [13] the aromatic amino acids in wool form a red azo dye with diazotized sulfanilic acid.

Such amino acids can only be found in the spindle cell layer below the scales, i.e. dyeing only occurs if the scale layer (cuticle) has been attacked.

Tyrosine is an aromatic amino acid which can be detected both in wool and in silk [14]. The dyeing reaction takes place as follows:

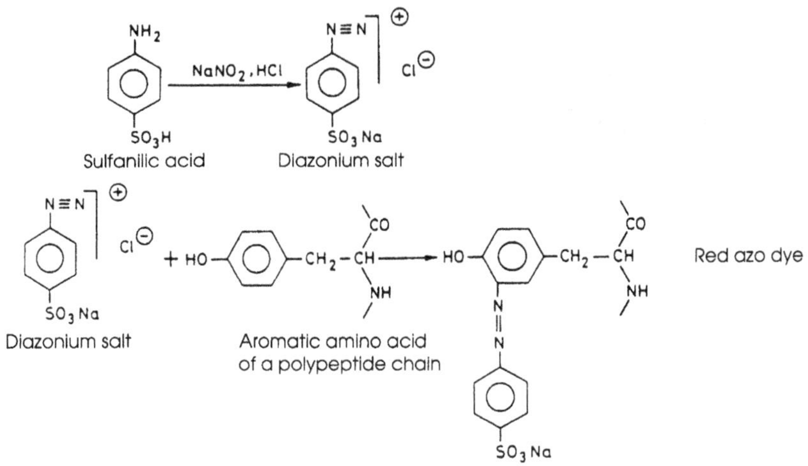

Substances and apparatus required for the production of diazonium salt:

1. Sulfanilic acid p. a.,
2. Sodium nitrite,
3. Conc. hydrochloric acid,
4. Sodium carbonate,
5. Glass filter crucible, plate diameter 30 mm, pore width 50–90 μm,
6. Burette,
7. Ice.

2 g of sulfanilic acid p. a. are suspended in 3 ml of dist. water, 2 ml of conc. hydrochloric acid are added and slowly diazotized (10–15 minutes) through drop by drop addition (burette) of a solution of 1 g sodium nitrite in 2 ml of dist. water.

The precipitated diazobenzene sulphonic acid is filtered through a G2 glass filter; it is then dissolved by adding 50 ml of a 10% sodium carbonate solution. When totally dissolved, 50 ml of dist. water and/or ice cubes are added. Because of the instability of the diazotate solution the temperature of the solution must not exceed 5 °C. The solution has to be used immediately after preparation. The dist. water used for the preparation of the diazotate solution as well as the sodium nitrite solution and the sodium carbonate solution has to be ice-cooled.

During the reaction, the wettability of the wool sample must be guaranteed. Therefore, the sample should be immersed in dist. water before performing the reaction. After centrifuging or squeezing, the sample is treated for 10 minutes in the ice-cooled diazotate solution and rinsed with ice-cooled dist. water. After

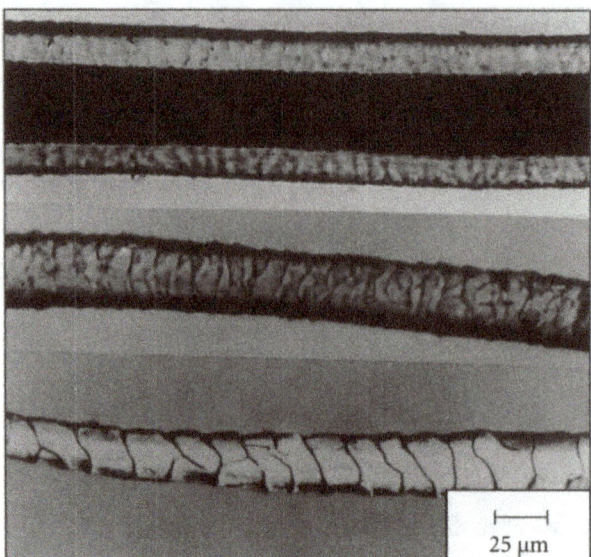

Fig. 26. Above: coarse wool fiber with medulla. Middle: Fine wool fiber. Below: Film imprint of a fine wool fiber.

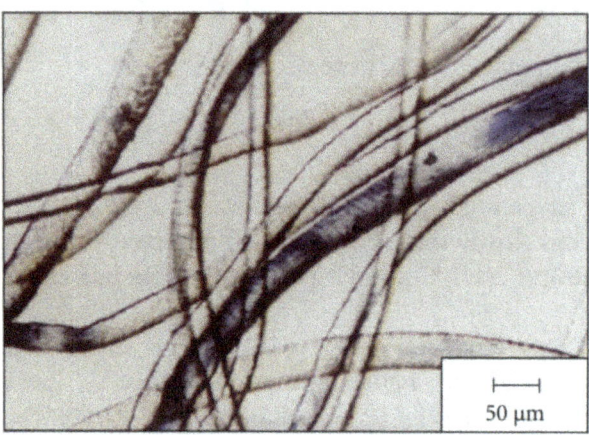

Fig. 27. Wool fibers with damaged places, stained with Cotton Blue-lactophenol which indicates all types of fiber damage.

Undamaged

Fig. 28. Wool samples, stained with Neocarmin W. The dye reagent mainly responds to badly alkaline-damaged wool.

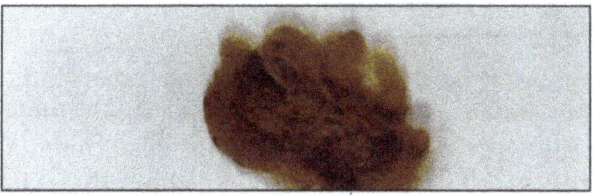

Alkaline-damaged

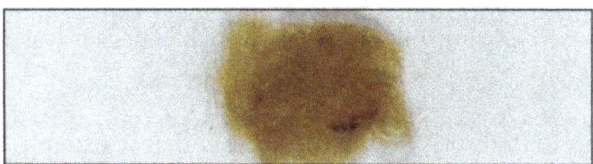

Acid-damaged

25 µm

Fig. 29. Wool fibers, treated with Pauly reagent, which, depending on the intensity of the damage, dyes damaged wool fibers yellow, orange or reddish brown. The dye reagent indicates all types of fiber damage.

further centrifuging or squeezing, it is either dried in the air or in the drying cabinet at 60 °C. The sample must always be thoroughly rinsed.

Since decomposition products can also dye completely intact wool fibers, it is important that the reaction – as already emphasized – is carried out quickly and that ice-cooled solutions are used. If experience is lacking, parallel control tests should be carried out with wool samples which are known to be intact or damaged. For comparison, damaged material can also be prepared, e.g. by boiling in a solution with 1 – 2 g sodium carbonate per liter. In this way, a reliable evaluation is guaranteed.

Fig. 30 – 31
Page 24
When evaluating this test method, it has to be remembered that there is no absolutely perfect wool without any damage. The wool of living sheep can be damaged through exposure to light, excrement, urine, alkaline dust, chemical influences or burs, Fig. 30, thistles and abrasion, Fig. 31. In all wool samples there are differing numbers of pre-damaged wool fibers, especially since further damage can be caused during raw wool scouring. Comparisons with untreated material are therefore highly recommended in order to determine which damage was caused during finishing and which damage was already present.

Fig. 32 – 33
Page 24 – 25
An important advantage of the Pauly reaction is the fact that it also responds to dyed wool. Figure 32 illustrates a green-dyed wool fiber. The green color of the damaged fibers was covered by the Pauly reaction, i.e. dyed wool does not interfere with the reaction. With a black-dyed material, however, recognition of the color reaction is more difficult. Staining with the Pauly reagent is usually only recognizable in the outer regions, as can be seen in Fig. 33; it shows a black-dyed wool fiber with a club-shaped thickening, the hair bulb.

Fig. 34
Page 25
The Pauly reaction is also quite helpful in fiber-analytical tests. Figure 34 shows wool fibers blended with synthetic protein fibers (e.g. casein fibers) after treatment with the Pauly reagent. The synthetic protein fibers are now dyed reddish brown. This reaction cannot be observed in any other man-made fiber. Attention must be paid to ensure that such fibers are not confused with damaged wool (see e.g. Fig. 31).

Fig. 35
Page 25
The Pauly reagent has also proven effective for the detection of skin particles. Skin particles, which are particularly found in skin wool or tanner's wool, are dyed reddish brown with Pauly reagent as is shown in Fig. 35 for rabbit hairs. A hair bulb dyed red with Pauly reagent can be seen on the lower right hand side. The pieces dyed reddish brown are skin particles. After the dyeing process these skin particles can create a very unpleasant

effect in wool yarns or textile fabrics by turning into dark dyed, nep-like soilings.

Other animal fibers or hair, e.g. rabbit hair, camel hair, mohair or alpaca display the same chemical reactions as wool. Because of the hair's internal structure (medulla, color pigments) and the fact that the scales are more compact compared to wool the cuticle scale structure cannot be recognized as such or only with difficulty, Fig. 35–40. Examination of the scale structure of these fibers is only possible with the aid of surface imprints.

Fig. 36–40
Page 26–28

2.1.2 Alkaline Damage to Wool and Wool-Like Natural Fibers

Fibers with sulphur-containing protein compounds are very sensitive to alkalies. This applies especially to heat treatments. For this reason, alkaline damage is found particularly frequently in wool materials. The Pauly reagent produces the orange to reddish brown dye already described. This dyeing is not specific for alkaline damage since the type of damage can be detected with Pauly reagent. However, alkaline damage is characterized by a stronger swelling of the damaged fibers, missing scale structure or splayed scales, curled fibers and the so-called crosier form, Fig. 40–42.

Fig. 41–42
Page 28

2.1.2.1 Dye Unlevelness in Wool Tops Made of Alkaline-Damaged Wool
– Practical Example

After dyeing wool tops with acid dyes, the combed tops were observed to be dyed differently although the same dye recipe had been used. In order to determine the reason for this, undyed top samples were examined for fiber damage with the Pauly reagent. This revealed that the Pauly reagent also dyed the top samples with different intensity, i. e. they contained a larger or smaller number of pre-damaged fibers, Fig. 43. Examination of the individual fibers under the microscope led to the conclusion that this was alkaline damage. Some fibers were strongly curled, Fig. 44, which is typical of alkaline damage. An inspection of the pH values showed that the combed top with most alkaline damage had a pH value of 9.3 while a pH value of 7.2 was found for the other top samples. Probably the rinse after alkaline scouring was not carried out with the necessary care. Originally, a mistake in the dyehouse was considered to be the cause of unlevelness. However, further tests proved this to be unjustified, because alkaline damage was identified. In the dyeing process alkaline-damaged wool fibers are generally dyed darker than the undamaged fibers [15].

Fig. 43–44
Page 29

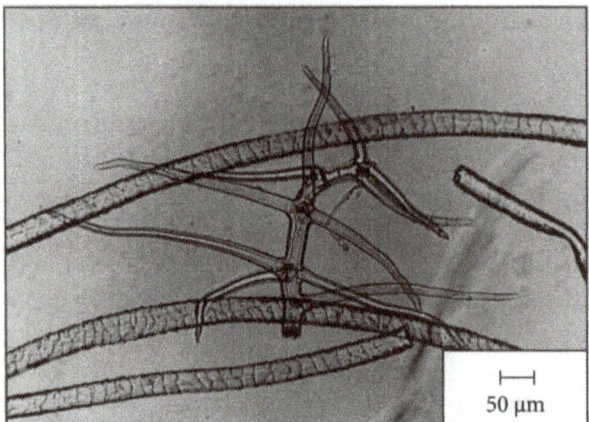

Fig. 30. Wool fibers with residues of vegetable constituents which can cause mechanical damage to the wool fibers.

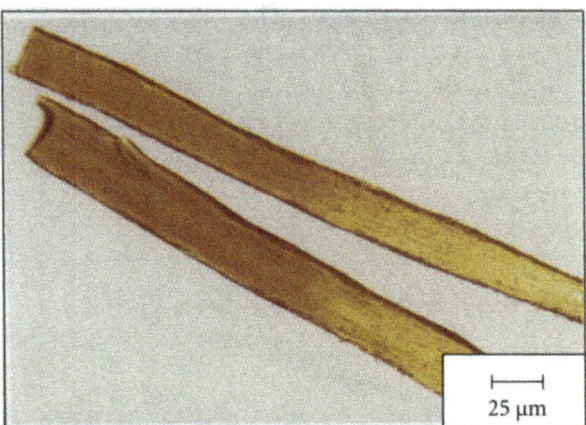

Fig. 31. Belly wool with abraded scales, dyed with Pauly reagent.

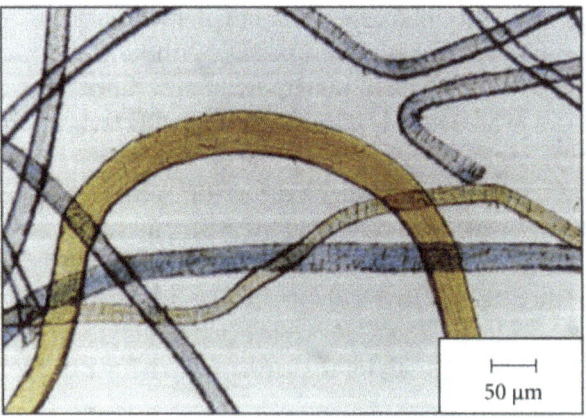

Fig. 32. Damaged fibers from a green-dyed wool, over-dyed with Pauly reagent. The reaction can also be used for dyed fabric.

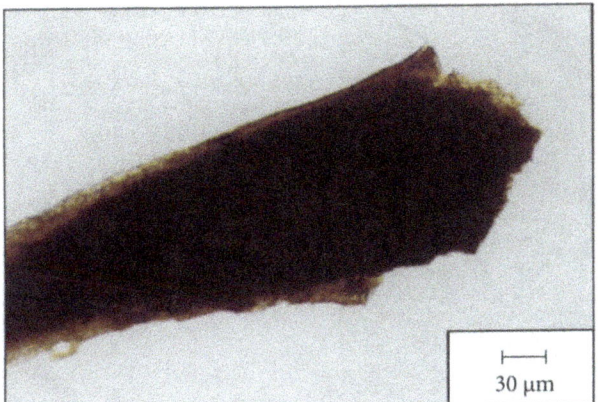

Fig. 33. Black-dyed wool fiber with hair bulb, dyed with Pauly reagent.

30 μm

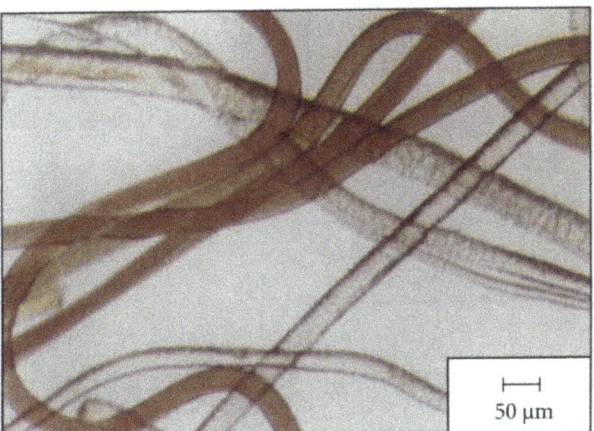

Fig. 34. Wool with synthetic protein fibers, dyed with Pauly reagent.

50 μm

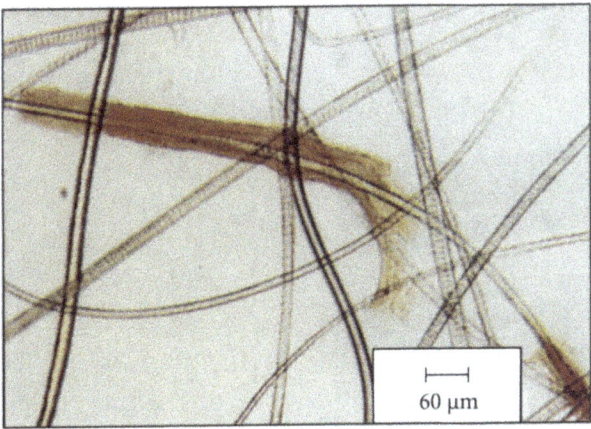

Fig. 35. Rabbit hair with hair root and skin particles which are dyed intensively reddish brown by the Pauly reagent.

60 μm

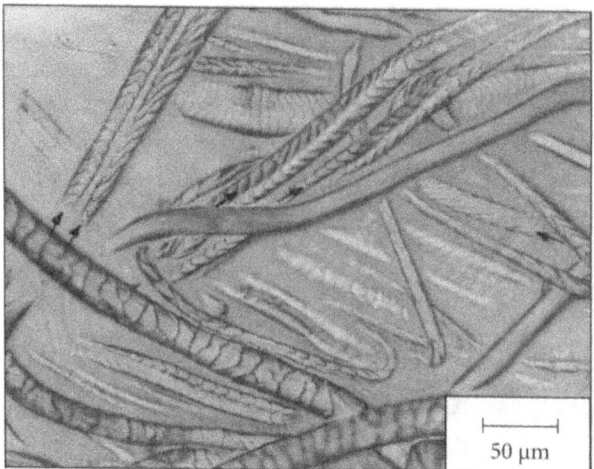

Fig. 36. Strongly magnified film imprint of an angora wool yarn (see arrows), wool and polyester. The internal structure of the fibers cannot be recognized on the imprint.

50 μm

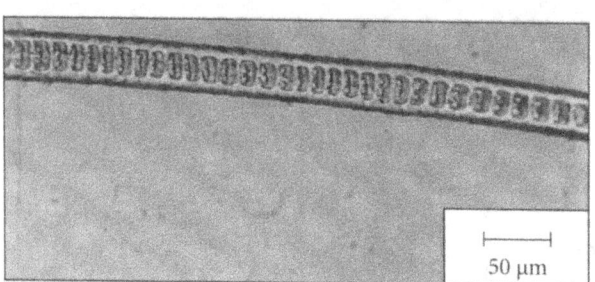

Fig. 37. Fiber sample of an angora rabbit hair with the typical internal structure in the form of rectangular cells.

50 μm

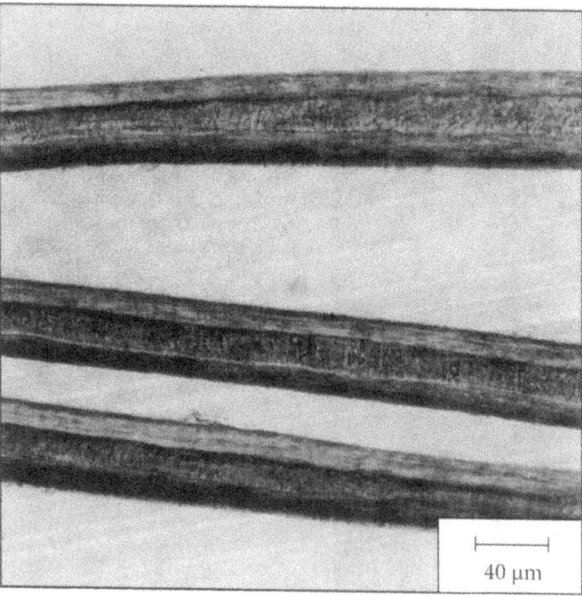

Fig. 38. Camel hairs, fiber sample. The internal structure (medulla) can be recognized easily.

40 μm

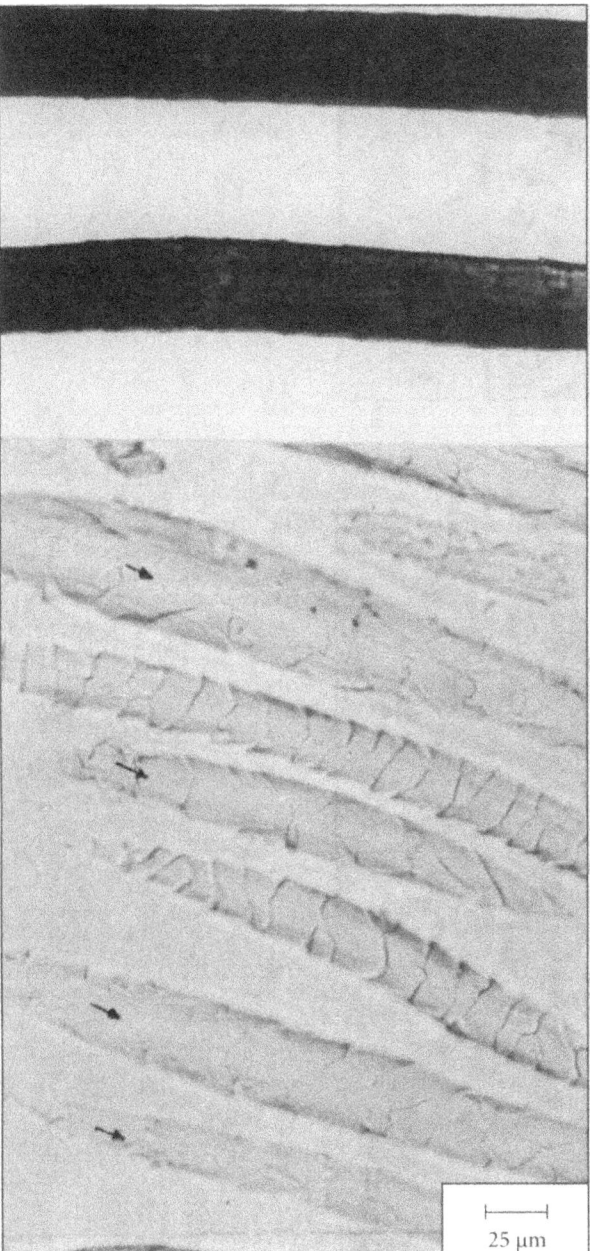

Fig. 39. Above: Fiber sample of deep-dyed mohair fibers. The surface structure can hardly be recognized. Below: Film imprint of wool and mohair fibers (arrows). The scales of the mohair fibers lie flatter, are larger, pointed and partially serrated.

25 µm

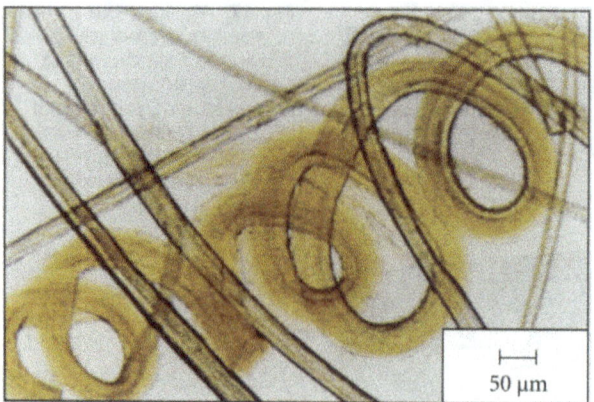

Fig. 40. Alpaca with a badly alkaline-damaged fiber which is swollen and strongly curled. A distinct dye formation occurs with the Pauly reagent.

50 μm

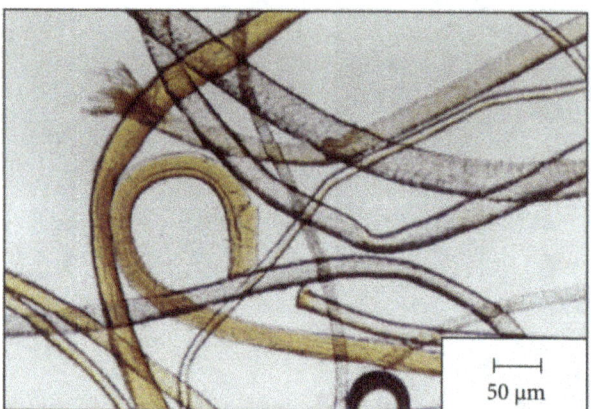

Fig. 41. Alkaline-damaged, gray-dyed wool, treated with Pauly reagent. The crosier form is typical of alkaline damage.

50 μm

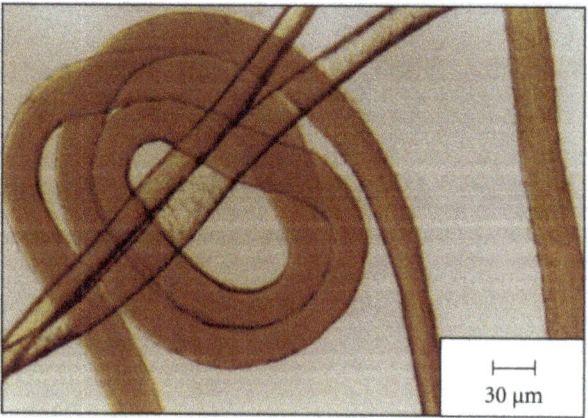

Fig. 42. Alkaline-damaged wool, dyed with Pauly reagent. One fiber has swollen badly and is curled up, also typical of alkaline damage.

30 μm

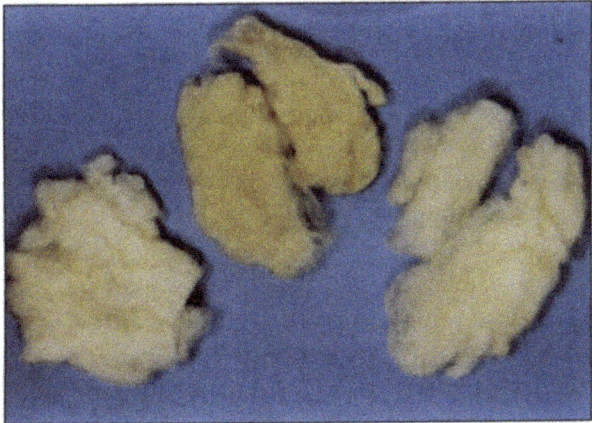

Fig. 43. Samples of wool tops, dyed with Pauly reagent. The middle sample is dyed more strongly than the other two and thus contains more pre-damaged fibers.

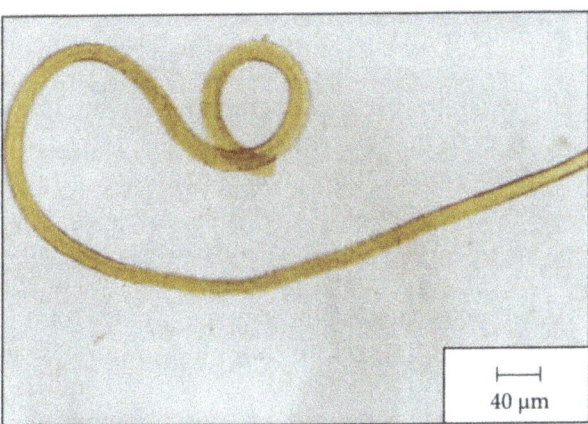

Fig. 44. Wool fibers from an alkaline-reacting wool top, dyed with Pauly reagent. The curly fiber form is typical of alkaline damage.

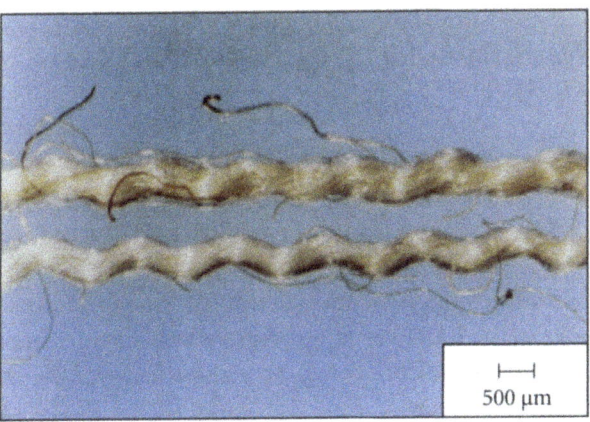

Fig. 45. Yarns from PES/wool, treated with Pauly reagent. The upper yarn sample contains more damaged wool fibers than the lower one.

2.1.2.2 Strength Loss in Wool, Caused by Alkalinely Reacting Untreated Yarn - Practical Example

The following example reveals the importance of supplying the institute or laboratory which has to examine the damaged fiber with specific information about the finishing of a fiber; for this purpose, attention must be paid to the starting material, i. e. the untreated material.

A wool yarn displayed a large loss in strength after the finishing process. The Pauly reaction and microscopic examination unambiguously proved alkaline damage. This could only be noticed after dyeing. For the dyer, the result was unintelligible since he had not utilized alkali during the finishing process. However, after the untreated yarn had been examined and details of the finishing were known, the correlations could be identified.

Before the dyeing process, the untreated yarns were crabbed in boiling hot liquors, to which only a small amount of fatty alcohol sulfate had been added. This treatment was simulated under laboratory conditions. Here a large loss in tensile strength could be registered in the yarns after crabbing and not just after dyeing. In addition, microscopic examinations showed that the untreated yarn already contained alkaline-damaged wool; this damage was much stronger in the crabbed yarn. The untreated yarn had a pH value of 9.5 – 10.0, proving clearly that an alkaline reaction had taken place. At low treatment temperatures, this pH value exerts only a slightly negative effect on wool, but in boiling hot liquors it accounted for the detected alkaline damage as well as for the decrease in tensile strength. As a result of these findings, the addition of acetic acid and a non-ionic surfactant to the crabbing bath was recommended. This surfactant can also develop its full wetting and scouring power in an acid treatment bath, thus eliminating the cause of the damage.

2.1.2.3 Dye Stains on Polyester/Wool Due to Local Alkaline Damage to the Wool - Practical Example

Fig. 45
Page 29

After dyeing in a light color, stains could be seen on fabrics made of polyester/wool. Figure 45 illustrates two yarn samples which were treated with Pauly reagent. They were taken from one spotted and one faultless section of the fabric. As can be seen in the figure, the thread taken from a spotted section shows more damage than a thread from an intact section of the fabric; this was proven by means of staining with Pauly reagent. The curled fiber forms also hinted at alkaline damage. Even in the case of very pale shades, wool with alkaline damage is dyed darker.

2.1.2.4 Limits of the Pauly Reaction – Strong Alkaline Damage

The depth of dyeing with Pauly reagent usually increases with the degree of damage. However, experience shows that the depth of dyeing decreases after exceeding a specific damage degree, see Fig. 46. This figure illustrates wool fibers with very bad alkaline damage, treated with Pauly reagent. The fibrous substance should be dyed reddish brown if the depth of dyeing increased constantly with the degree of damage. However, the damaged fibers are partially yellow. Therefore, one should never rely on dyeing alone but should always combine it with microscopic examination. A decrease in the color intensity is probably due to the fact that the aromatic amino acids are separated through dissolution from the spindle cell layer as a result of strong alkali action; thus they can no longer participate in the reaction.

Fig. 46
Page 32

2.1.3 Acid Damage to Wool

Although wool fiber is rather resistant to acid action, it sustains considerable damage during inappropriate carbonization or long periods of boiling in strongly acid dyeing liquors. Here it must be noted that the resistance of alkaline-damaged wool to acid is reduced.

The carbonizing process can give rise to many problems. Uneven squeezing, suctioning and centrifuging as well as uneven drying of the acidified pieces with different acid concentrations lead to local damage to the wool.

Slight acid damage cannot always be identified directly through simple microscopic examination. The scale layer may in principle be well preserved, while the less resistant fiber interior may already be damaged. The Pauly reaction simplifies the recognition of damage and provides a good idea of the degree of acid attack. Wool becomes brittle as a result of stronger acid action. It is then very sensitive to mechanical stress such as rubbing and abrasion. Under these circumstances mechanical stress almost always leads to damage, i.e. splitting and breaking, of the wool fibers. Sections of wool fibers are split into spindle cells, Fig. 47. They show fibrillar dissolution phenomena. Incident light is scattered diffusely at the split fibers. This produces lighter areas and stains and/or gives the impression of lighter deposits. On acid-damaged wool reserving effects can occur in areas of stronger acid-induced attack. Even with small carbonization differences, the usual acid dyes are unable to produce even dye shades. Therefore, after carbonization it is recommended to use chrome complex dyes which are most suitable to compensate for carbonization differences. However, chrome complex dyes are of no help in the case of stronger acid-

Fig. 47
Page 32

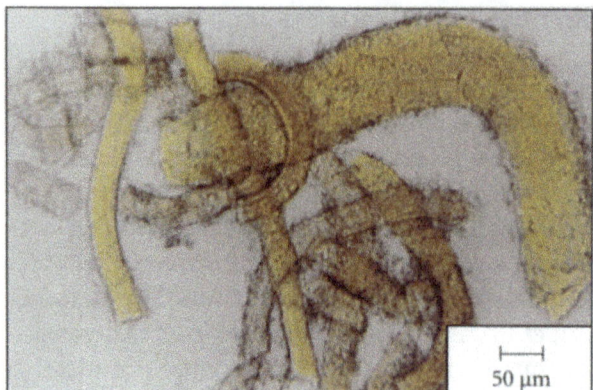

Fig. 46. Badly alkaline-damaged wool fibers, dyed with Pauly reagent.
In spite of strong damage to the wool fiber fragments, some fibers are not dyed reddish brown but only yellow.

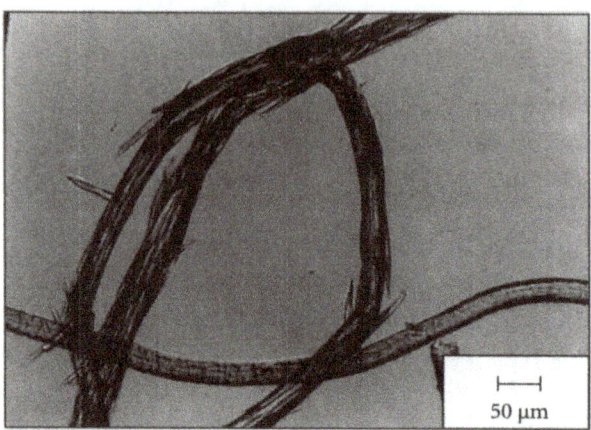

Fig. 47. Acid-damaged fibers dissolved into spindle cells.

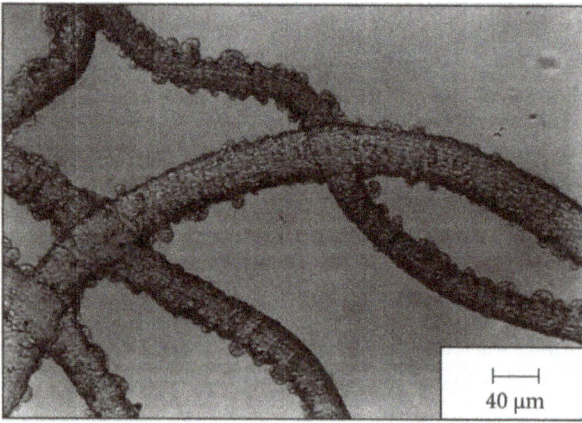

Fig. 48. Detection of acid-damaged wool fibers by means of the KV reaction. In the case of acid-damaged wool the bubbles already form after 1–2 minutes.

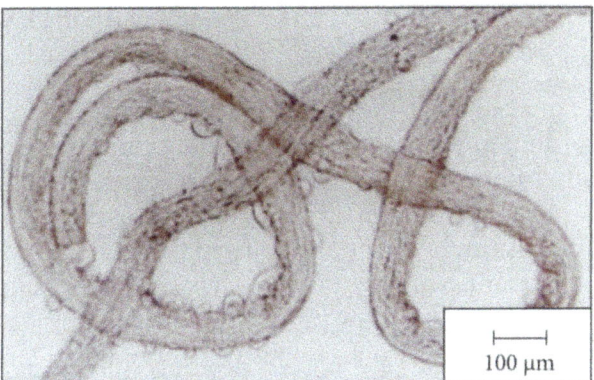

Fig. 49. Wool fibers, treated with Pauly reagent; subsequently acid damage was detected by means of the KV reaction.

Not chlorinated

Fig. 50. Wool, chlorinated and unchlorinated, dyed with Neocarmin W.

Neutrally chlorinated

Acid-chlorinated

induced attack, i. e. fibrillar dissolution of the wool fiber. The darker the dyeing of the material, the greater the effect of the damage.

Fig. 48
Page 32

For the detection of acid-damaged wool, the swelling reaction with ammonium potassium hydroxide according to Krais and Viertel (KV reaction), has proven to be very useful [16]. With this method acid-damaged wool can be detected in the initial stage; several wool fibers on a microscope slide are covered with a cover glass, ammoniacal potassium hydroxide (produced from 20 g of caustic potash + 50 ml of conc. ammonia, under cautious shaking and cooling) is poured over the slide. After 1 to 2 minutes rapidly growing bubble-shaped blisters are formed, Fig. 48. On intact wool the blisters can only be detected after about 10 minutes and there are far fewer than in the case of acid-damaged wool. With alkaline damaged wool, the bubbles only form after about 30 minutes, if at all.

Fig. 49
Page 33

Depending on the individual case, it can be advantageous to combine the Pauly reaction with the KV reaction by drying the wool sample stained with Pauly reagent and then carrying out the KV reaction under a microscope [17]. This simplifies the separation of the intact fibers from the damaged fibers for microscopic examination, Fig. 49.

2.1.4 Chlorine Damage to Wool

The chlorination of wool in order to improve its luster, produce a non-felting finish and increase its substantivity in the production of printed fabric, is no simple operation. This process influences the macromolecular fiber composition and the structure of the wool fiber. Chlorination breaks down or completely removes the scale layer. A complete removal occurs particularly if the fabric is chlorinated in the acid pH range – the scale layer is then virtually separated. Through chlorination, wool becomes more lustrous (silk wool), displays more or less strong yellowing phenomena and can occasionally lose its wear resistance. However, the affinity of chlorinated wool for dyes as well as the wettability are increased, which is essential for wool printing.

Fig. 50
Page 33

The detection of chlorine damage under the microscope is quite complicated, since, just like any damaged wool, wool with chlorine damage is generally dyed orange to reddish brown through the reaction with Pauly reagent. In the case of natural white or light-colored wool, chlorinated and unchlorinated wool can be distinguished by means of dyeing with Neocarmin W. The wool samples are treated with the Neocarmin W solution for 5 minutes at room temperature and then rinsed under running water until the water is clear. Wool which was lightly chlorinated in a neutral and/or slightly alkaline liquor is dyed orange, wool chlorinated in a strong acid liquor is dyed brown. Unchlorinated wool is dyed yellow, Fig. 50.

2.1.4.1 Chlorination of a Wool Carpet (Gold Afghan), Increase of Luster, Gold Effects – Practical Example

Several wool pile tufts from a Gold Afghan were examined; the dark green base of the tuft turned into a golden-brown color towards the top. It had to be examined how the golden-brown shade and the luster had been produced.

For microscopic examination of the scale structure, imprints of the pile tufts were produced on polypropylene films. Under the microscope it could be seen that the wool fibers displayed a relatively well developed scale structure at the base of the tuft; at the top of the tuft the scale structure was hardly recognizable or could not be detected at all, Fig. 51. Apparently, during washing the carpet had been treated with chlorine. This had two effects:

Fig. 51
Page 36

1) The shade changed into a golden-brown.

2) The luster of the carpet was increased as a result of extensive destruction of the scale layer.

In order to achieve the gold effect and the luster, the chemical structure at the top of the wool tufts was changed deliberately (no accidental and undesired damage of wool).

2.1.4.2 Chlorination of Wool in the Non-Felting Finish

In the non-felting finish of wool, e.g. in the Hercosett process, the scale edges are etched away and the wettability is increased through cautious chlorination; the scales are then masked with a polymer. This compensates for the pawl effect of the scales, Fig. 52, which plays an important role in the felting of wool. Figure 53 illustrates the imprint of two wool fibers as a photomontage of yarn imprints of one treated and one untreated wool yarn.

Fig. 52–53
Page 36

The non-felting finish of wool can also be distinguished from pure chlorine treatments through comparision of the following dyeing reactions:

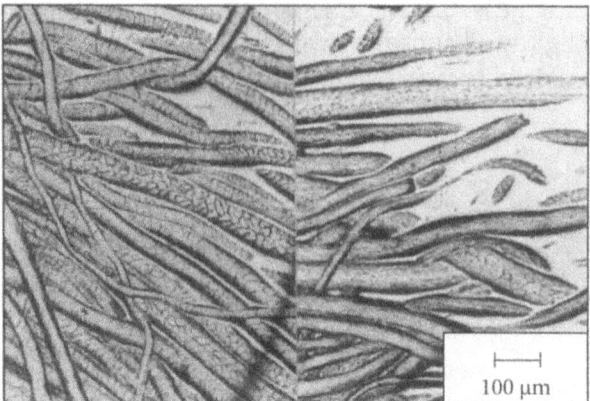

Fig. 51. Left: Film imprint of the base of a wool pile tuft from a Gold Afghan. The scale structure of the wool fibers can be clearly recognized. Right: Film imprint of the top of a wool tuft from a Gold Afghan. The wool fibers no longer display a scale structure. Most of the scales were removed during chlorination.

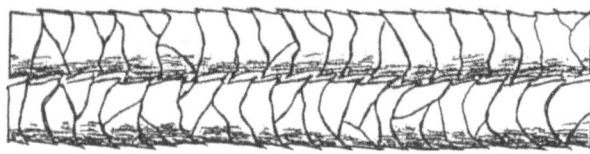

Fig. 52. Pawl effect of the scales which plays an important role in the felting of wool, schematical.

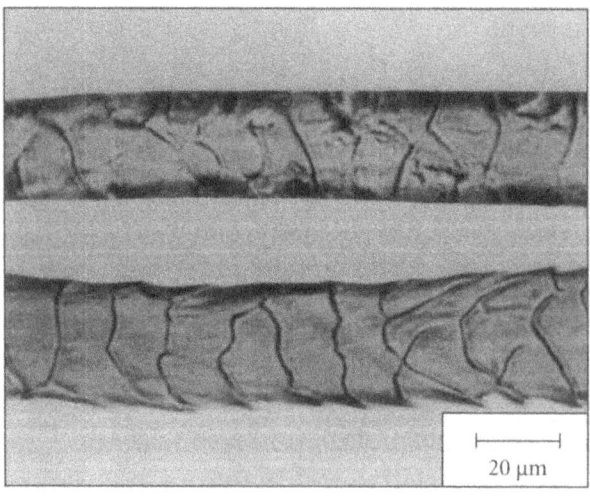

Fig. 53. Film imprint of wool fibers. Above: Wool with a non-felting finish. Scale edges etched off by chlorination, masked with a polymer. Below: Untreated wool with a good surface structure and slightly splayed scales.

a) **Neocarmin W**
Untreated wool: Dyed yellow.
Wool with non-felting finish: Dyed brown.

b) **Pauly reaction**
Untreated wool: Not dyed.
Wool with non-felting finish: Dyed orange to red (similar to damaged wool)

c) **Acid dye combination [18]**
Untreated wool: Dyed yellow.
Wool with non-felting finish: Dyed blue.
Chlorinated wool: Dyed dirty green.

Staining with an acid dye combination is carried out with:

0.8% by weight Supramin Yellow GW (Bayer) (C.I. Acid Yellow 61),
1.0% by weight Acilan Fast Navy Blue R (Bayer) (C.I. Acid Blue 92),
5.0% by weight Glauber's salt,
6.0 m/l Acetic acid (60%),
2.0 g/l Sodium acetate,

Temperature: 40 °C,
Liquor ratio: 1 : 40,
Treatment time: 10 minutes.

2.2 Chemical Damage to Silk

2.2.1 Chemical Composition, Structure and Microscopy of Silk

Fiber damage to silk [19] can only be perceived if the chemical and microscopic structure of silk is known in detail.

Like wool, silk is a protein fiber. However, in contrast to wool it contains only sulfur-free amino acids. Two types of natural silk can be distinguished:

a) Mulberry silk (Bombyx mori); "genuine silk"

b) "Wild silk", its best known representative being Tussah.

Mulberry silkworms feed on the leaves of the mulberry tree.

The raw silk thread, i.e. the cocoon thread, of the Bombyx mori consists of the external cover, the so-called silk gum (sericin) and the internal thread (fibroin), the actual silk substance usable as a textile thread. In contrast to the fibroin, the sericin layer is brittle and inelastic. It covers the beautiful luster of the fibroin.

Fig. 54
Page 39

Under the microscope one recognizes that the raw silk thread (cocoon thread) consists of two single filaments which are surrounded by the gum, Fig. 54. The gum is cracked, split and irregularly attached to the fibroin filaments. There are many areas in which the cocoon thread is already split.

Fig. 55–56
Page 39

Degummed silk, in contrast, has independent single filaments, Fig. 55. Their longitudinal view is smooth and structureless. The width of the individual fibers varies. Their cross-section is triangular, the corners are slightly rounded, Fig. 56. Sometimes the silk fibers display a slight longitudinal streakiness, which provides a hint of the fibrillar structure of the silk. In mulberry silk the proportion of silk gum is 20 – 25 %, in a few exceptional cases – Thai silk – it can be as much as 38 %. The gum can be golden-yellow, yellowish, yellowish brown or yellowish green, depending on its origin. Occasionally it may be white.

Fig. 57–58
Page 40

The best known representative of wild silk, Tussah, is produced by Tussah silk worms, i.e. by worms of the genus Antheraea which feed on oak leaves. Even after degumming it remains brown. Therefore this silk has to be bleached with hydrogen peroxide before dyeing. The brown color is probably due to the sericin content of the fibroin filaments. The removable gum content in Tussah silk is 14 – 17 %. Under the microscope Tussah silk is unevenly thick and band-shaped; it has fine vertical stripes which can be attributed to the fibrillar structure of the fibers. Compared to mulberry silk, the threads of Tussah silk are flatter and considerably wider, Fig. 57. Typically Tussah silk has flattened spots, so-called squeezed spots. They form at the intersection of two silk threads overlapping in the cocoon. The fiber cross-section of Tussah silk is wedge-shaped and angular; similar to the longitudinal view, it shows a flat, band-shaped structure, Fig. 58. The band-shaped constitution of the slightly twisted fibers lends a strange luster to the Tussah.

Compared to mulberry silk, Tussah silk is more resistant to alkalies and acids. Therefore, higher pH values can be used during degumming.

Degumming is an essential processing step in order to achieve level dyeing and printing. Degumming of silk was and is an expensive process. In former times soap or enzymes were used. Soap degumming of silk takes place in mildly alkaline soap baths for a period of 3 1/2 to 4 hours at almost boiling point, thus separating the sericin. The pH value must not exceed 9 – 10. Higher pH values accelerate degumming but can cause alkaline damage to the fibroin. With the synthetic degumming agent Miltopan SE (Cognis) degumming takes 40 to 60 minutes. The product contains a fiber protective agent for the fibroin so that higher pH values can be used than during soap degumming. When using Miltopan SE for degumming, the fiber protective agent allows a pH value of

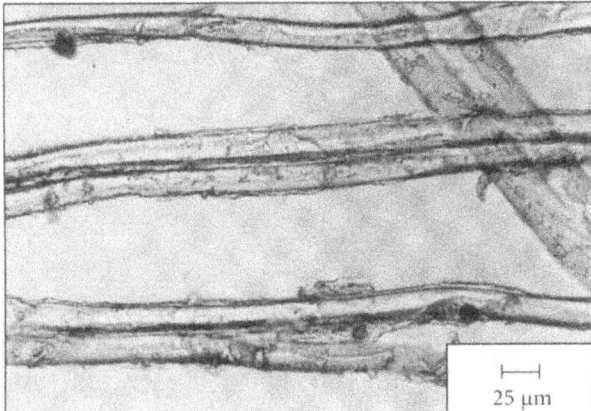

Fig. 54. Raw silk of the Bombyx mori (mulberry silk).

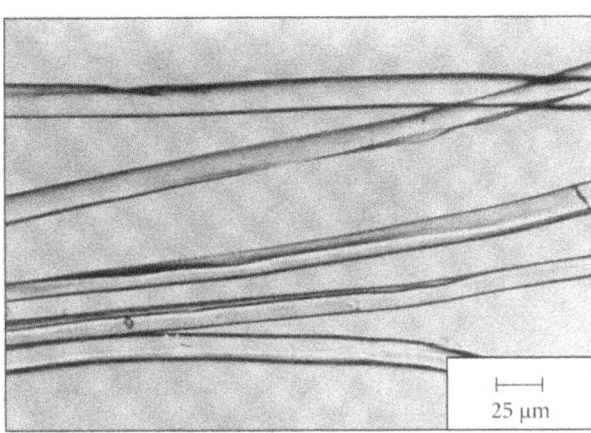

Fig. 55. Mulberry silk, degummed.

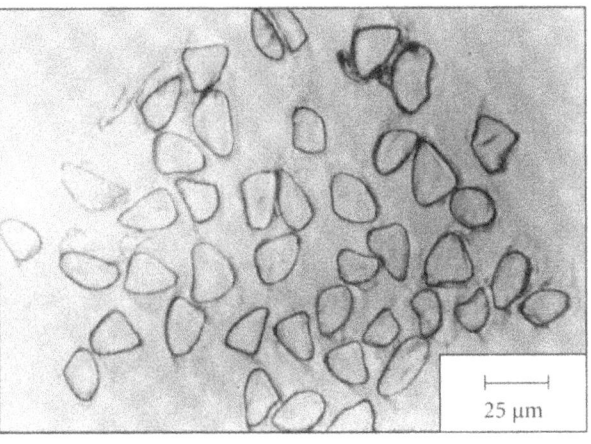

Fig. 56. Fiber cross-section of mulberry silk, degummed.

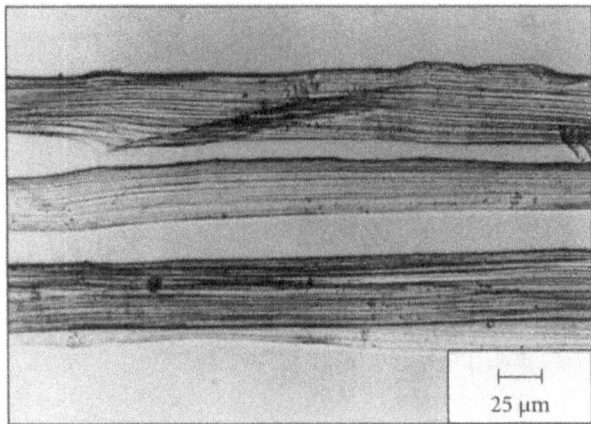

Fig. 57. Tussah silk, fiber sample.

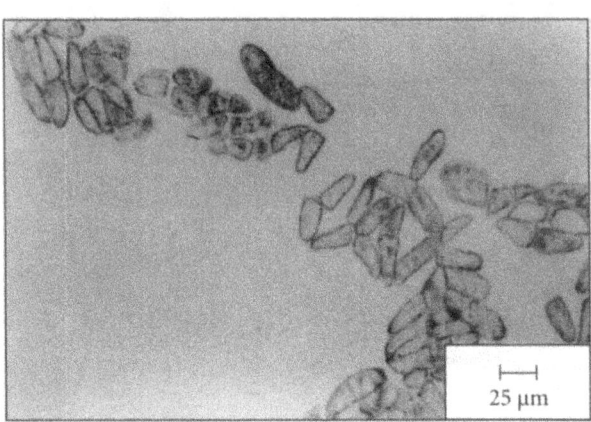

Fig. 58. Fiber cross-section of Tussah silk.

11–11.5, without causing alkaline damage to the fibroin. This is advantageous because the pH value can be adjusted with sodium hydroxide instead of sodium carbonate which must be used in soap degumming. If the prescribed pH values are exceeded during degumming, the silk can easily be damaged because the liquors are too alkaline. The resulting alkaline damage is the most frequently observed chemical damage to silk.

2.2.2 Detection of Chemical Damage to Silk with Pauly Reagent

As in the case of wool, Pauly reagent can be used for the detection of damage to silk [20]. However, detection is more difficult because the gum is also dyed orange to red by Pauly reagent. Therefore, this reagent can also be used to determine whether the silk is degummed or not. If the reagent dyes the silk yellow, it is degummed and undamaged. If the silk turns orange-red, it must be determined microscopically whether the silk is degummed. Only if no gum can be detected does the orange-red color indicate chemical damage to the silk (see chapter 2.2.3). Degummed Tussah silk is dyed yellowish brown by the Pauly reagent. Bleached Tussah silk turns slightly orange since Tussah is slightly damaged during bleaching. Mechanically damaged silk fabric is dyed red by the Pauly reagent (see chapter 3.2).

For these reasons, the practical application of the reagent presupposes a certain experience, especially since diazobenzenesulfonic acid – as emphasized above – is easily decomposed. The resulting decomposition products can also dye completely intact silk fibers.

In order to avoid drawing wrong conclusions, the reaction must therefore be carried out rapidly; the solution must be cooled with ice, and the samples to be tested must remain in the reaction solution for no longer than 1–2 minutes, because, in contrast to wool, silk has no scale layer but only a thin fibroin skin. With a longer residence time, especially if the temperature is too high, the reagent could penetrate the coating of the intact silk and enter the fibril layer. If experience is limited, it is advisable to carry out parallel tests with silk samples which are known to be intact and/or damaged. One can also produce damaged material for comparative tests, e.g. through soap degumming at pH values over 10, in order to guarantee a sound evaluation.

Figure 59 illustrates the cross-section of a cocoon thread and how the individual layers have been dyed with Pauly reagent. Figure 60 shows the corresponding dyeing of mulberry silk in the raw and degummed state as well as after alkaline damage.

Fig. 59–60
Page 44–45

2.2.3 Control of the Degumming Effect

The safest method to control degumming is the analytical determination of the residual gum content. For this purpose, a silk sample of about 10 g is conditioned in the standard atmosphere at 20 °C and 65 % relative atmospheric humidity and weighed on an analytical balance (precision of measurement 0.1 mg). The sample is then degummed as follows under optimal conditions with a liquor ratio of 1: 40 – 1: 60:

12 g/l Miltopan SE (Cognis),
pH value 11.0 – 11.5 (if necessary this can be adjusted with sodium hydroxide)
40 minutes at 94 °C.

The sample is then rinsed with hot and cold water, centrifuged, dried at 105 °C and weighed again in the standard climate after conditioning. The weight loss, as a percentage of the weight-in quantity, is equal to the gum content.

Fig. 61–62
Page 46
A quick way to check on degumming is the staining test with Neocarmin W; unfortunately, it is not very reliable. According to the specifications of the manufacturer, degummed mulberry silk is dyed golden-yellow by this dye. However, our own experience has shown that the gum turns blue, brown, violet or reddish violet, depending on the origin of the mulberry silk. Degummed silk is dyed blue-violet but the depth of staining is not as deep compared to raw silk. Figure 61 illustrates a cross-section of a raw silk fabric before, and Fig. 62 after dyeing with Neocarmin W. With the aid of the staining process, gum can also be recognized in the fabric cross-section.

The staining test with the substantive dye Sirius Red F3B 200 % (Bayer) (C.I. Direct Red 80) yields a better result. The silk is immersed into a 1 % dye solution for one minute and then rinsed:

Silk with gum: Red,
Silk without gum: Undyed.

In the case of Tussah silk, the degumming effect can hardly be evaluated via a staining test with Neocarmin W, since the fibroin filaments still contain sericin constituents. Therefore, raw as well as degummed Tussah silk is dyed olive.

Under the microscope, however, degummed and raw silk can be distinguished by means of the Cuoxam test [21]. After the silk fiber has been immersed in Cuoxam (see appendix), it can be seen how the fibroin is slowly dissolved while the sericin remains insoluble.

Another option for microscopic degumming control is the preparation of surface imprints. Figure 63 illustrates the imprint of a raw silk fabric on a polypropylene film. The bonding of the silk threads with gum can be recognized
clearly on the imprint. Figure 64 shows the imprint of the same fabric after
degumming. It can be seen clearly that the bonds are now dissolved.

Fig. 63 – 64
Page 46 – 47

2.3 Chemical Damage to Cotton

In the finishing of cotton [22] damage is most frequently caused during bleaching. For the bleaching of cotton, hydrogen peroxide is the most important oxi-
dative bleaching agent, both in the semicontinuous and continuous processes,
and in a discontinuous process with a high liquor ratio. Heat and activating
additives, e.g. sodium hydroxide, enhance the bleaching effect [23]. In order
that the process does not run out of control, stabilizers are added to the bleaching liquor; the stabilizers can be inorganic or organic (sodium silicate and/or
Stabilol types, Henkel) and counteract the self-decomposition of the hydrogen
peroxide. These additives lead to an equilibrium between activation and stabi-
lization of the bleaching liquor. The equilibrium can be disturbed by catalysts,
e.g. heavy metal ions, thus enhancing activation. In the presence of catalysts
this effect leads to an overoxidation of the cellulose which reduces its strength.
This in turn produces decomposition products which are called oxycellulose,
not to be confused with hydrocellulose which can be attributed to damage to
the cellulose from mineral acids; hydrocellulose partially forms as a result of
acid splashes, e.g. if the fabric is acidified with mineral acids after alkaline processes. However, acid damage is very rare because acidification is usually carried out with organic acids, e.g. acetic acid or formic acid.

2.3.1 Microscopy of Cotton

Under the microscope, cotton has band-shaped fibers which display characteristic twists and distortions. It is therefore easily distinguished from all other
fibers, Fig. 65. The lumen of cotton has a different width, depending on the
degree of maturity, and is often hard to recognize in the longitudinal view.

Fig. 65
Page 47

However, man-made fibers can also display cotton-like twists. If there is any
doubt, the fibers should be immersed in zinc chloride-iodine solution (Merck).
This reagent dyes all cellulosic fibers blue. If the fibers are dyed blue, the fabric
is definitely cotton and not a synthetic fiber. A disturbance of the reaction with
zinc chloride-iodine solution from resin finishes can be recognized easily.
Directly after immersing in zinc chloride-iodine solution the cotton fibers

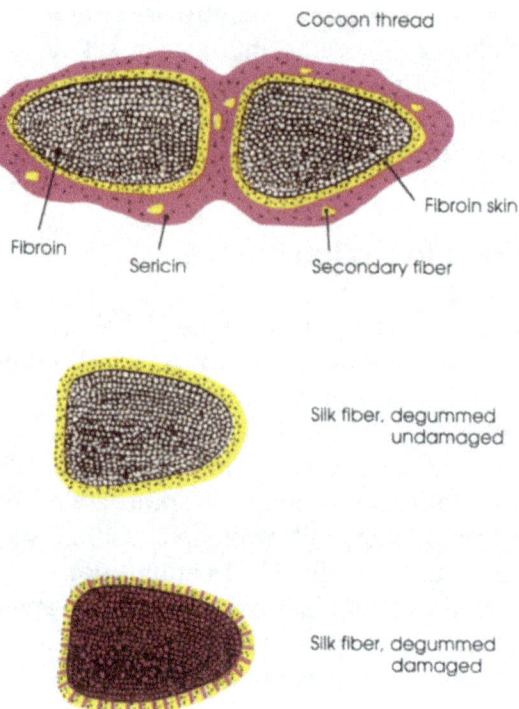

Cocoon thread

Fibroin skin

Fibroin

Sericin

Secondary fiber

Silk fiber, degummed undamaged

Silk fiber, degummed damaged

Fig. 59 A. Cross-sections of a strongly magnified mulberry silk cocoon thread as well as an intact and a damaged silk fiber in the degummed state, after dyeing with Pauly reagent.

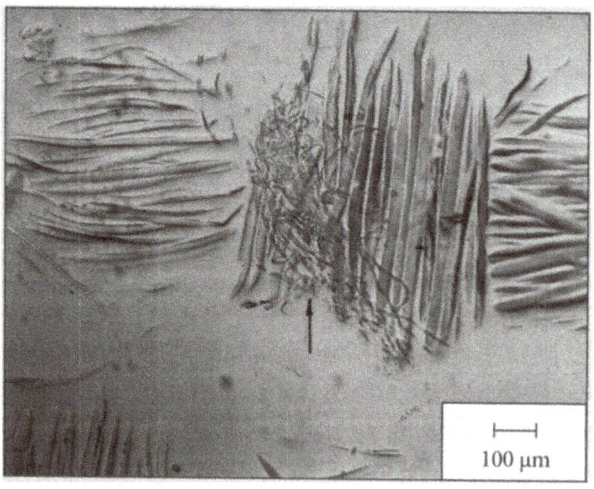

100 μm

Fig. 59 B. Film imprint of degummed silk fabric with silk lice, caused by agglomeration of secondary strands.

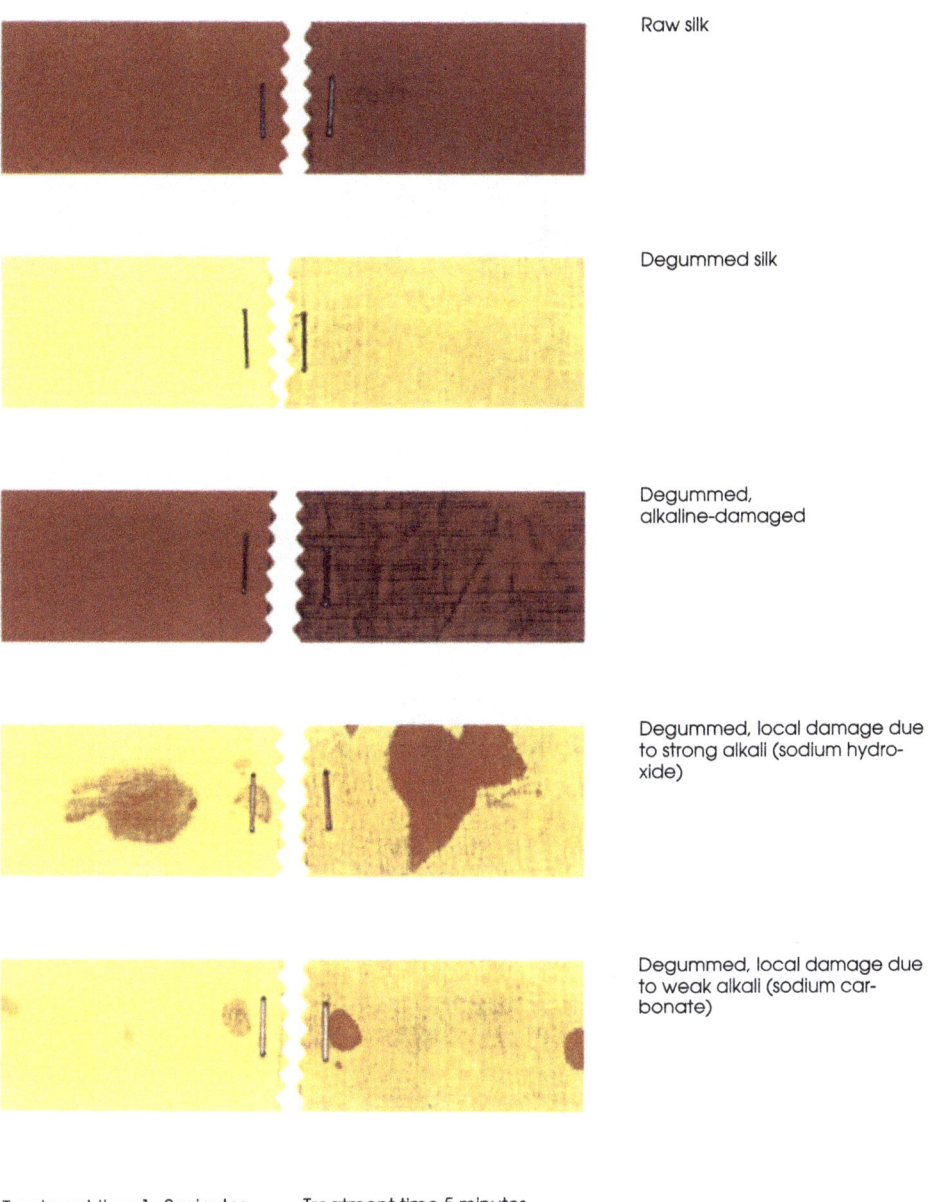

Raw silk

Degummed silk

Degummed,
alkaline-damaged

Degummed, local damage due
to strong alkali (sodium hydroxide)

Degummed, local damage due
to weak alkali (sodium carbonate)

Treatment time 1–2 minutes   Treatment time 5 minutes

Fig. 60. Detection of damage to natural silk with Pauly reagent.

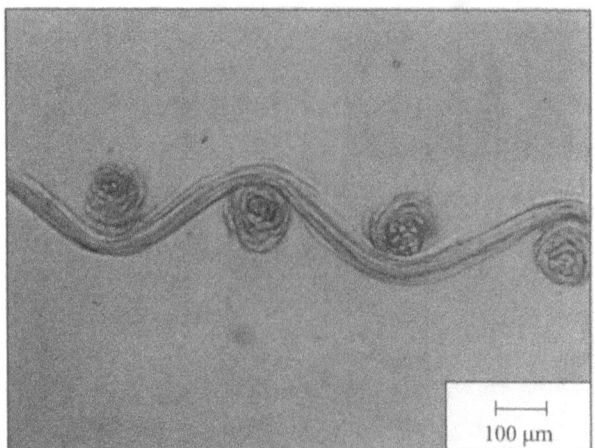

Fig. 61. Cross-section of a raw silk fabric.

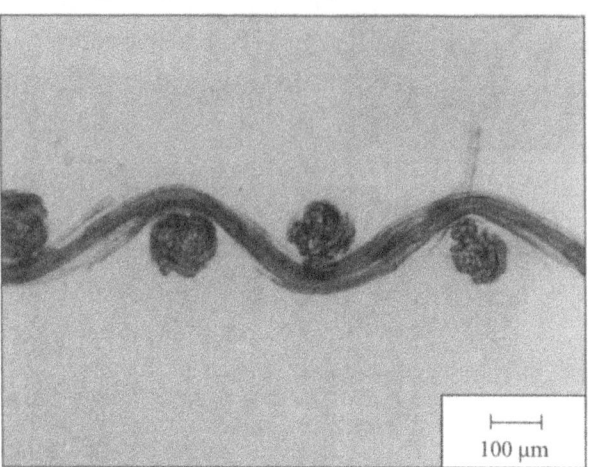

Fig. 62. Cross-section of a raw silk fabric. Gum dyed with Neocarmin W.

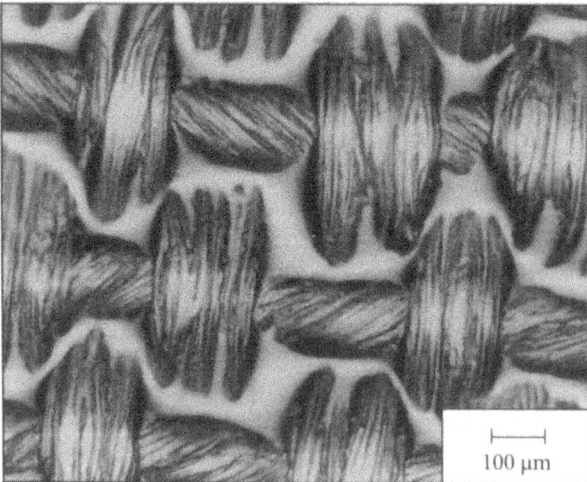

Fig. 63. Film imprint of a raw silk fabric. The silk fibers are bonded by the gum.

Fig. 64. Degummed silk fabric as a film imprint.

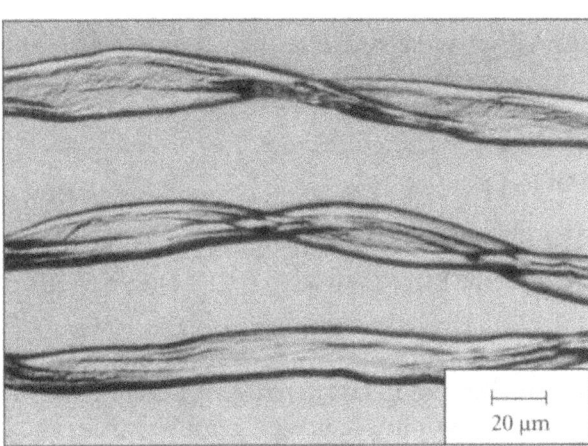

Fig. 65. Cotton fibers with typical twists and distortions.

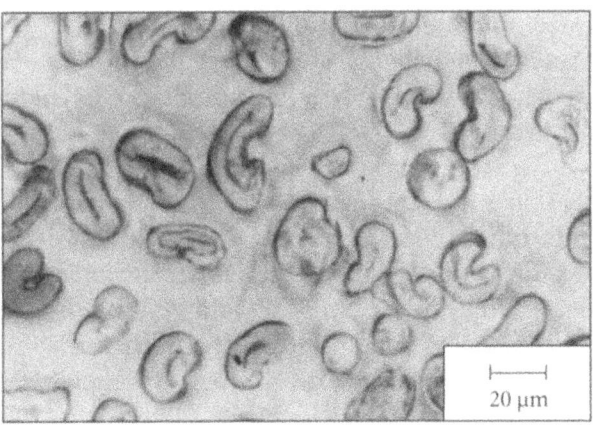

Fig. 66. Fiber cross-section of raw cotton.

finished with synthetic resins are dyed yellow but after a short time the color changes to blue. Another way to detect cellulosic fibers is the Cuoxam test. Cuoxam dissolves all cellulosic fibres; the mature, raw cotton fibers form very typical, barrel-shaped and/or spherical expansions (ballooning, cuticle ring) before dissolution [2].

Fig. 66
Page 47

Fiber cross-sections of cotton, Fig. 66, can vary greatly, depending on the degree of maturity of the fiber. Mature, raw cotton has a bean-shaped, oblong, oval cross-section. Immature cotton fibers have an oblong and flat cross-section, in dead fibers it is narrow and flat with a broad lumen.

Fig. 67
Page 49

The degree of maturity of the raw cotton can be determined with the aid of a simple dyeing test. With the so-called red/green test [24] mature cotton (thick cell wall) is dyed red and immature cotton (thin cell wall) is dyed green. Figure 67 illustrates the color test for two cotton card slivers. The green-dyed carded sliver with a large proportion of immature fibers was less suitable for spinning than the red-dyed carded sliver with the mature fibers. Dead cotton – prematurely damped offfibers – is not, or only slightly, dyed green with this dye reagent.

The recipe for the red/green test is:

1.2 % by weight Diphenyl Red® 5B 182 % (Ciba-Geigy) (C.I. Direct Red 81),
2.8 % by weight Solophenyl Green® BL (Ciba-Geigy) (C.I. Direct Green 27),
Liquor ratio 1 : 40 (dist. water).

The sample is thoroughly wetted in boiling dist. water and then treated for 15 minutes in a dyeing bath at boiling point. It is then removed from the bath; 2.5 % by weight of sodium chloride is added, the sample is left to dye for another 15 minutes and again removed. An additional 2.5 % by weight of sodium chloride is added and the dyeing process is continued for 15 minutes. The sample is rinsed twice in cold water, and then treated for 30 seconds in boiling water with a liquor ratio of 1 : 40; the sample is removed, rinsed twice in cold water and dried in the air after squeezing.

Fig. 68
Page 49

The dead cotton fibers only have a thin cell wall; they are completely transparent which leads to optically lighter dyeing. Furthermore, dead cotton fibers do not display the windings and twistings which are typical of normal, mature cotton. Instead they display wrinkles and foldings. Black-dyed cotton fibers are illustrated in Fig. 68. It can be seen that the mature cotton fibers are dyed completely while the dead fibers are apparently not dyed.

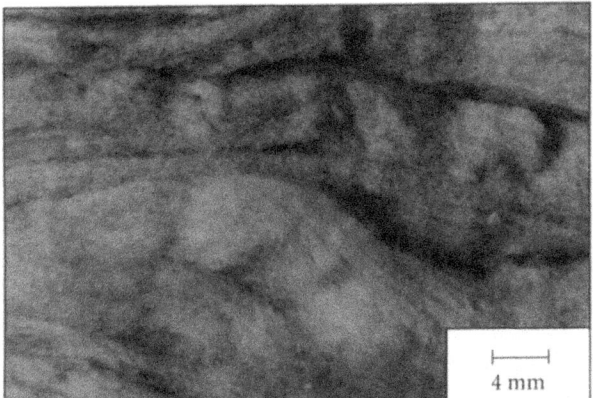

Fig. 67. Red/green test on two cotton card slivers. Mature cotton fibers: Red. Immature cotton fibers: Green.

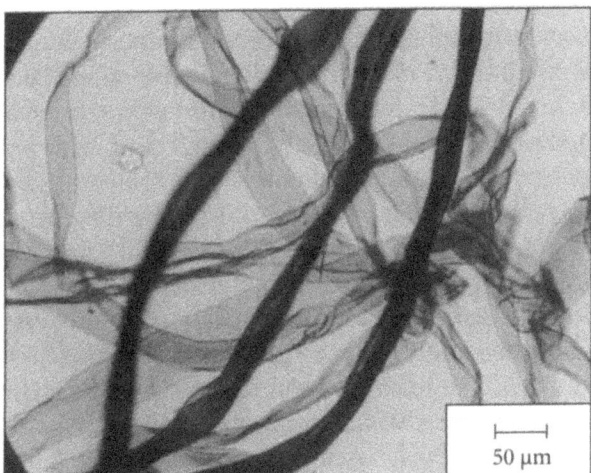

Fig. 68. Cotton fibers dyed black. The mature cotton fibers are dyed evenly throughout while the dead cotton fibers are completely transparent due to their thin cell wall.

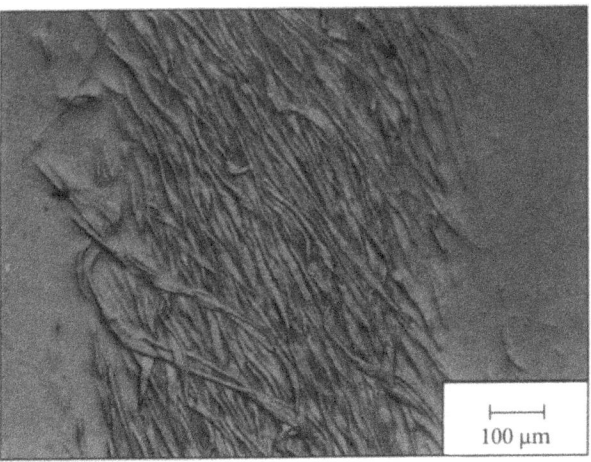

Fig. 69. Film imprint of an unmercerized cotton yarn, which can be recognized from the twists and screws of the fibers.

Fig. 69–70
Page 49/52 During mercerizing, the twists in the cotton fibers untwist. Under the microscope they are structureless, smooth and/or roller-shaped. The bean-shaped fiber cross-section of the mature fibers changes into a round form. With yarn imprints, mercerized and non-mercerized cotton can be easily distinguished. In this case, fiber samples or fiber cross-sections are not necessary. In the case of non-mercerized yarns the yarn imprint illustrates that all fibers have the typical twists and windings, whereas in mercerized products most fibers have a smooth, roller-like structure, Fig. 69 and 70. Furthermore, under the same conditions mercerized cotton is dyed far more intensively dark blue by zinc chloride-iodine solution than normal cotton.

2.3.2 Detection of Oxycellulose and Hydrocellulose

Fig. 71
Page 52 To detect chemical damage to cotton, both staining reactions and microscopic reactions have been proposed. Oxycarmin (Fesago, Heidelberg) and Fehling's solution [25] are the best known and acknowledged staining reagents available. The alkali treatment swelling reaction (pinhead reaction) according to Koch [26] is suitable as a microscopic detection reaction. Figure 71 illustrates locally damaged cotton fabrics treated with Fehling's solution and/or Oxycarmin.

Oxycarmin dyes oxycellulose dark blue to dark violet. Hydrocellulose remains undyed. Fehling's solution imparts a reddish brown color both to oxycellulose and hydrocellulose. The reaction is based on the reducing characteristics of chemically degraded cellulose. The depth of dyeing and/or the shade differences between oxycellulose and hydrocellulose which can be seen in Fig. 71 resulted accidentally. The dye shades could easily be the other way round. They depend on the degree of damage. These micrographs could be interpreted in such a way that it should be quite easy to distinguish undamaged cotton, oxycellulose and hydrocellulose by means of staining reactions according to the system in table 2.

Table 2. Staining reactions for undamaged cotton, oxycellulose and hydrocellulose

Fiber	Oxycarmin test	Fehling's solution
Oxycellulose	positive	positive
Hydrocellulose	negative	positive
Undamaged cotton	negative	negative

Fig. 72
Page 52 However, tests have revealed that staining reactions only work perfectly if freshly damaged material is used that has not undergone alkaline aftertreatment –

through dyeing or aftersoaping processes [15]. Thus the staining reactions are not always reliable. For this reason, these reactions should always be combined with microscopic detection reactions.

For microscopic examination of chemically damaged cotton, the previously mentioned pinhead reaction has proven to be very effective. In this reaction a sodium hydroxide solution (15 %) is used. If undamaged cotton fibers are immersed in the solution, pinhead swellings protrude from the cut ends which must be cut cleanly with a pair of sharp scissors or a razor blade (no squeezed fiber ends). They are formed through small increases in diameter when the interior of the fibers protrudes from the cut ends, due to the fact that the shell of the cotton fiber swells less than the fiber interior, Fig. 72. However, the reaction only works on mature cotton. As has already been emphasized, immature fibers, with thinner walls compared to mature cotton, do not protrude from the interior through the broad cell channel since, due to the larger lumen, there is sufficient room to spread inside. While the mature cotton fibers lose their twists and take a cylindrical, smooth structure when immersed in 15 % sodium hydroxide (as during mercerisation), the immature cotton fibers form new, regular twists (Fig. 72, center). Thus, via this reaction, immature and mature fibers can be distinguished under the microscope.

Fig. 72
Page 52

Dead cotton fibers show no pinhead reaction. Therefore, for the evaluation of the pinhead reaction immature and dead cotton fibers must not be taken into consideration. This complicates the examination, particularly if there is a large proportion of immature and dead fibers in the yarn or textile fabrics.

In chemically damaged cotton, the pinheads and/or rivet heads become flatter and flatter, depending on the degree of fiber damage. In the case of obvious damage, there are no rivet heads and/or no pinheads protruding from the fiber ends; they remain flat.

This effect must be attributed to damage of the outer skin and hence the fibers only swell in diameter. If chemical damage is very strong, apart from swelling in diameter, notches and splits can be detected. According to Koch, to some extent these splits represent a rough image of the reduction in chain length of the cellulose molecules, Fig. 73. Such fiber images are frequently found with cotton damaged by mineral acids. However, one cannot simply infer acid damage because the same fiber images had already been observed with over-oxidized cotton bleached with chlorine, Fig. 74.

Fig. 73–74
Page 54

In short, it can be said that the pinhead reaction is a relatively safe method for the detection of chemical damage to cotton. However, the pinhead reaction

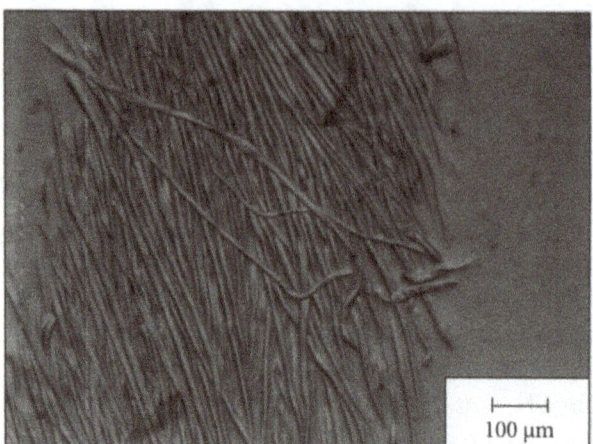

Fig. 70. Film imprint of a mercerized cotton yarn. The fibers are roller-shaped, structureless and smooth.

Fig. 71. Left: Hydrocellulose, detection with Fehling's solution. Middle: Oxycellulose, detection with Fehling's solution. Right: Oxycellulose, detection with Oxycarmin.

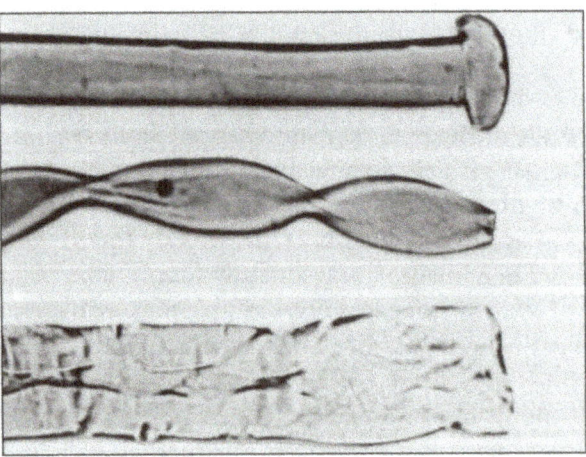

Fig. 72. Alkali treatment swelling reaction (Pinhead reaction). Above: Mature, undamaged cotton. Middle: Immature cotton. Below: Chemically damaged cotton.

does not permit a clear distinction between oxycellulose and hydrocellulose. For this purpose, further criteria must be used, e.g. the pH value in the damaged area. This can be checked by dripping Universal Indicator liquid (Merck) or by pressing pH paper moistened with dist. water onto the fabric. With the aid of Congo Red paper (Merck), the fabric can then be tested for mineral acids.

The discoloration of cotton in damaged areas is also of importance. A dark brownish discoloration indicates that the cotton has been carbonized. Providing that there are no iron deposits, this is an important hint that the damage was caused by mineral acids. Therefore, brownish colored damaged areas should first be tested for iron.

2.3.3 Bleaching Damage Due to Catalysts

All heavy metals such as iron, copper, chromium, nickel and manganese as well as their oxides and salts are catalysts which may cause bleaching damage; catalytical damage is most frequently caused by iron.

Nowadays, catalytical damage due to copper is quite rare although it is far more serious than that caused by iron [27]. Modern bleaching equipment for batch and continuous processes is made of stainless steel and has no copper mountings. With the exception of manganese, other heavy metals are only of theoretical interest as catalysts for bleaching damage. Iron can be contained as abraded metal in grease stains, often bound in oil and graphite. Rust particles, iron chips and/or iron dust can also be spun into the thread or woven during the manufacturing process. Iron ions can be contained in industrial water, e.g. when it dwells in iron pipes over the weekend. It is then contained as hydroxide in alkaline bleaching liquors. Iron and manganese can also be brought in with the raw cotton.

In the catalyst-containing areas of the fabric, hydrogen peroxide is spontaneously decomposed during the bleaching process; nascent oxygen is produced, which leads to the formation of oxycellulose. After bleaching this can be seen in the form of holes or cracks in the fabric. The higher the temperature and pH value, the greater the damage to the cellulose fibers in the metal-containing areas. However, catalytic fiber damage can also be caused during cold pad-batch bleaching, since, with exposure times of approx. 24 h, even at low temperatures, the cotton around the catalysts is overoxidized. Even if the damage is not strong enough to lead to the formation of holes, the overoxidation will result in uneven dyeing. Figure 75 illustrates both damage phenomena.

Fig. 75
Page 54

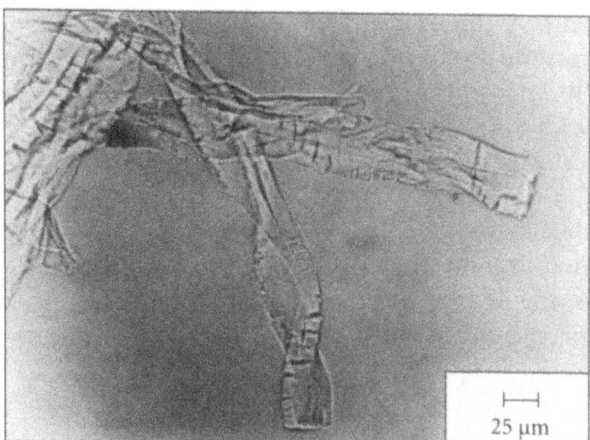

Fig. 73. Cotton badly damaged by mineral acid. Flat fiber ends, strong swelling in diameter, notches and splits after immersion in 15% sodium hydroxide.

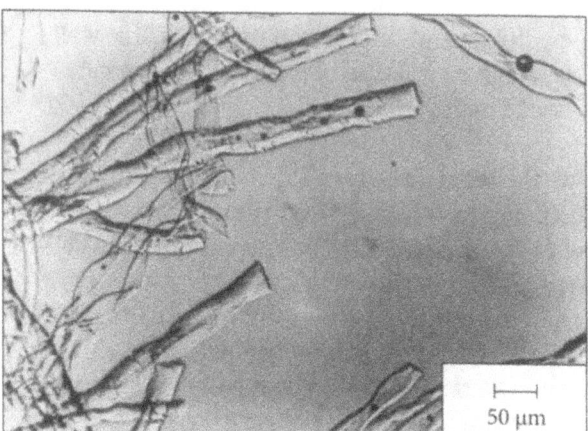

Fig. 74. Cotton damaged by overoxidation with chlorine. Flat fiber ends, notches and splits in the longitudinal direction after immersion in 15% sodium hydroxide.

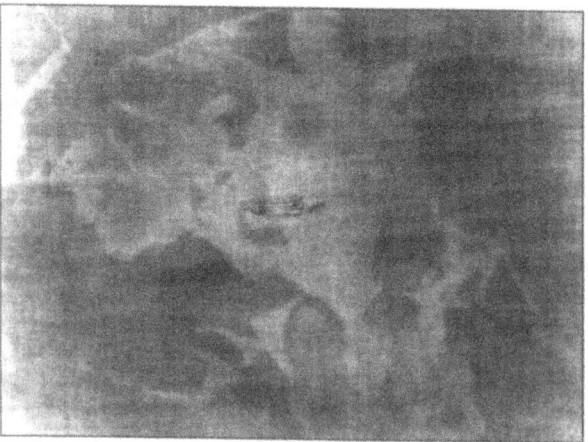

Fig. 75. Chemically damaged cotton. The brittle areas of the piece are dyed lighter than the undamaged parts.

Iron can be easily detected on cotton if it occurs in the form of inwoven rust particles, iron chips or filtered iron. Diluted hydrochloric acid p.a. is dabbed onto the respective areas and left there for a short period of time to react; 5% potassium cyanoferrate (II)-solution is then dripped onto the acidified areas of the fabric. In the presence of iron, an intensive blue color develops, Fig. 76, known as Prussian Blue. A 10% ammonium thiocyanate solution can also be utilized. The iron-containing areas are dyed intensively red with this reagent, Fig. 77. If iron is mixed with graphite and oil, iron impurities in the grease stains can only be detected with a magnifying glass or microscope. However, iron detection by means of the described methods only works if the iron ions are trivalent - this should normally be the case, since in the alkaline range bivalent iron is oxidized into trivalent iron by atmospheric oxygen. After bleaching with hydrogen peroxide, iron(III)-ions will always be present. If there is any doubt, these areas of the fabric should be treated with a strongly diluted hydrogen peroxide solution before they are tested for iron. In the case of very dark dye shades such as black, navy blue or dark red, iron cannot be reliably detected on the fabric by color reactions. Therefore, the sample is ashed, the ash is put into diluted hydrochloric acid (p.a.), and a potassium hexacyanoferrate (II)-solution or ammonium thiocyanate solution is added.

Fig. 76 - 77
Page 56

Examination of the fabric under UV light also provides hints of catalytic bleaching damage because the fluorescence of a bleached and optically brightened fabric is extinguished by metal impurities. The metal-containing spots on the fabric then appear as dull, dark, non-fluorescent areas under UV light.

In order to reliably prevent catalytic bleaching damage, it does not always suffice to add complexing agents to the alkaline pad liquors during semicontinuous and continuous processes. Fine iron dust impurities and iron chips can only be removed by special treatment with oxalic acid before bleaching. In most cases, grease stains with abraded metal are caused in the weaving mill. They should be eliminated by clean working practice. Damage to the fabric resulting from such impurities can be very serious.

2.3.4 Different Phenomena of Catalytic Bleaching Damage

In the majority of cases catalytic bleaching damage is locally restricted. Outside the range of the damaged areas, the fabric generally displays normal strength. Typically, with catalytic bleaching damage, either only the weft yarns or only the warp threads are destroyed, i.e. only one of the two fabric directions is affected. The damage always occurs on those threads which showed metal impurities before bleaching, Fig. 78.

Fig. 78
Page 56

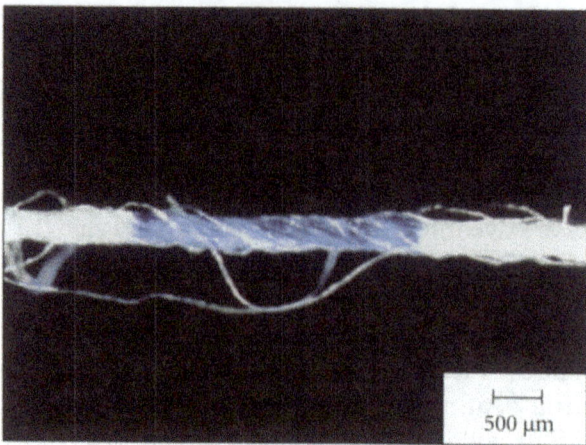

Fig. 76. Iron-containing grease stain on a cotton yarn. Iron detection with the Prussian Blue reaction.

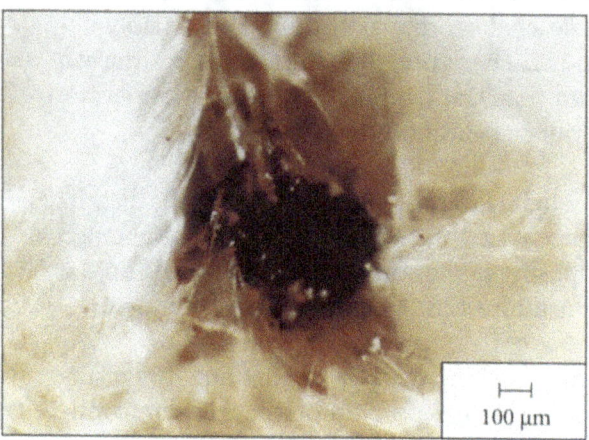

Fig. 77. Grease stain on cotton knitwear. Iron detection with ammonium thiocyanate.

Fig. 78. Fabric made of cotton and modal fibers with catalytic damage. Only the warp threads were destroyed after bleaching. The local limitation is typical of catalytic bleaching damage.

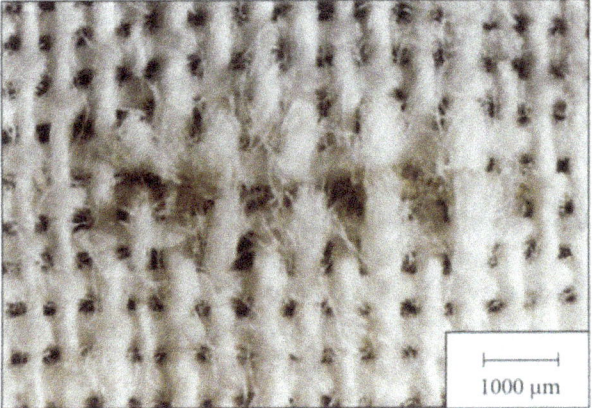

Fig. 79. Catalytically damaged, broken warp and weft yarns in a cotton fabric (warp direction from the left to the right).

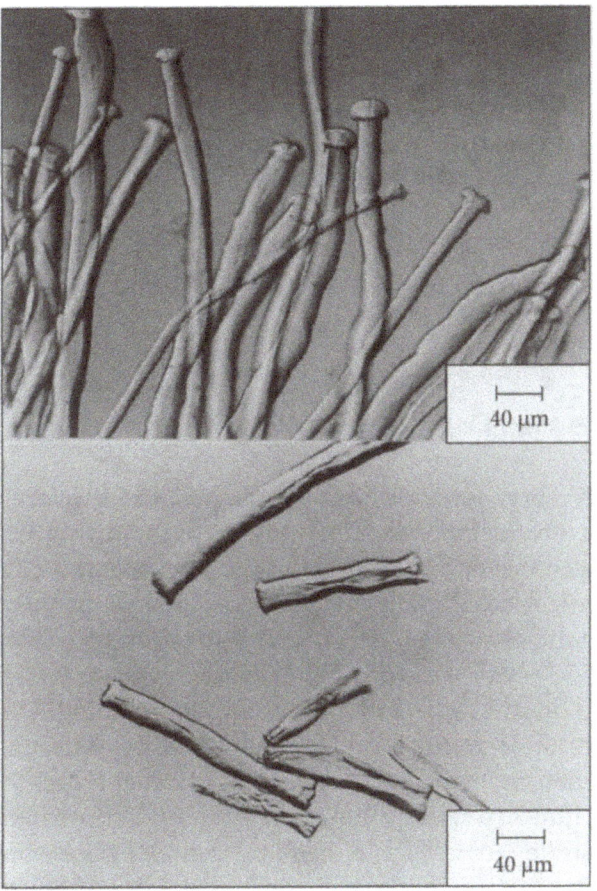

Fig. 80. Cotton fibers from a fabric with catalytic damage after bleaching. Above: Pinhead reaction outside the range of the damaged areas – positive; the cotton is not damaged. Below: Pinhead reaction in the range of the damaged areas – negative; cotton is clearly damaged.

2.3.4.1 Holes and Tears in a Bleached Cotton Fabric - Practical Example

Fig. 79 – 80
Page 57
After bleaching, a cotton fabric developed numerous holes which gave the impression of cuts and/or tears, Fig. 79. However, microscopic examination together with the pinhead reaction clearly showed that the fiber material was chemically damaged near the broken threads. Outside the damaged areas the material was absolutely intact because, after immersion in 15% sodium hydroxide, optimally developed pinheads and/or rivet heads developed, Fig. 80. Iron could not be detected in the damaged areas. As a corresponding piece of untreated material was available, bleaching tests with hydrogen peroxide were carried out under laboratory conditions. After the tests had been completed, dark grease stripes appeared which were Fig. 81
Page 60 approx. the same size as the resulting tears. Under mechanical strain in these areas, e. g. by pressing from below with a fingernail, the threads split. When the fabric was rubbed, the soiled yarn ripped. The Prussian Blue reaction and grease-soluble dyes revealed that the impurities were iron deposits and oil, Fig. 81.

These impurities only occurred on the warp threads. They had obviously been deposited on the fabric during weaving. Figure 79 illustrates that the iron-containing warp thread piece is broken off and that the wefts, which had been in contact with the iron impurities at the crossing points, are also broken.

Contact with iron-containing material alone is sufficient to cause catalytic bleaching damage, i.e. not every thread that breaks necessarily contains iron. This leaves the impression of a tear and/or a cut.

2.3.4.2 Small Holes in a Bleached Cotton Fabric – Practical Example

Fig. 82
Page 60
After bleaching, a cotton fabric developed numerous small holes. In the outer regions of the holes the Oxycarmin test was positive, thus proving that the cotton was locally damaged through overoxidation. The local restriction of the damage hinted at catalytic bleaching damage, but the catalyst, namely iron, could not be detected in the damaged areas. The iron-containing, badly damaged yarn pieces had obviously broken off. Slightly darker and smaller soilings could only be recognized on a few pieces of the fabric after bleaching. Traces of iron could be detected in these areas by means of the Prussian Blue reaction. Damage to cotton fiber material could also be detected here (flat pinheads after immersion in 15% sodium hydroxide); however, fiber material was not so brittle that mechanical strain would lead to the development of holes, Fig. 82. This practical example shows that it is not always simple to detect the cause of

damage, the catalyst. This is particularly valid if it is not possible to examine the corresponding untreated material.

2.4 Chemical Damage to Synthetics

Chemical damage to synthetics is much less frequent than damage to natural fibers because they are more resistant to chemical influences during finishing. The most frequently observed chemical damage to synthetics is acid damage to polyamide. It frequently occurs during dyeing of polyamide fabrics when conc. formic acid is added without dilution to the dyeing liquor. Other practice-related damage is alkaline damage to acetate filaments and – quite rarely – acid damage to polyester.

2.4.1 Acid Damage to Polyamide Knitwear - Practical Example

Light stains and brittle, hard areas could be seen on black-dyed polyamide knitwear, Fig. 83. They had presumably been caused by a permanent antistatic finish with a product based on alkaline cross-linkable polyglycol polyamines. Observations had been made that light stains developed during drying of the permanent antistatic finish. The fabric became brittle in these areas.

Fig. 83
Page 61

The microscopic examination, coupled with staining reactions, revealed that none of the specifically modified synthetic resins used for finishing could be detected in the respective areas.

If the light stains had actually been caused by such substances this would be visible, as can be seen in Fig. 84. The precipitated permanent antistatic finish would form flat pads on the fabric. In this case, they were dyed with Sirius Light Rubine (Bayer) or Sirius Pink BB 143 % (Bayer). This can be done in the following way: a solution with 1 g/l of dye is produced. The sample is put into the dye solution for 10 minutes at room temperature and subsequently rinsed under running water until the water is clear.

Fig. 84
Page 61

However, under microscopic examination, our practical example showed that the polyamide fibers were totally deformed in the light, hardened areas, Fig. 85. The fibers were largely dissolved and/or melted into lumps and corroded at the surface. This is typical of acid damage, but thermal damage could not be excluded. In case of doubt acid-damaged polyamide fibers can be detected by staining with a solution of 3 g/l Rhodamine B extra (Merck), 10 m/l 60 %

Fig. 85 – 86
Page 61/64

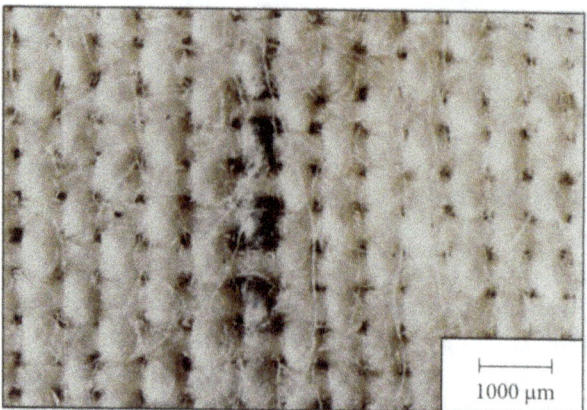

Fig. 81. Stains that contain abraded metal in a cotton warp, leading to catalytic bleaching damage. Iron detection with the Prussian Blue reaction.

1000 µm

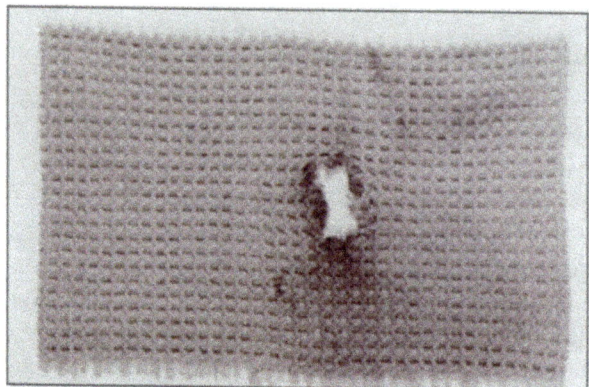

Fig. 82. Cotton fabrics with small holes after bleaching, typical of catalytic bleaching damage. Areas of damage with positive Oxycarmin test.

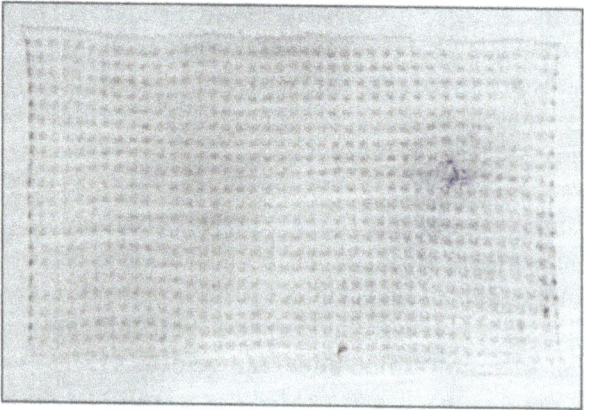

Positive iron detection with the Prussian Blue reaction on small grease stain.

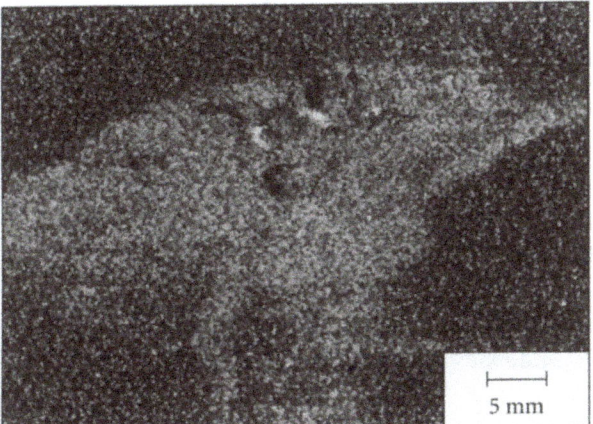

Fig. 83. Black-dyed polyamide knitwear with light stains and brittle areas which have resulted from acid splashes.

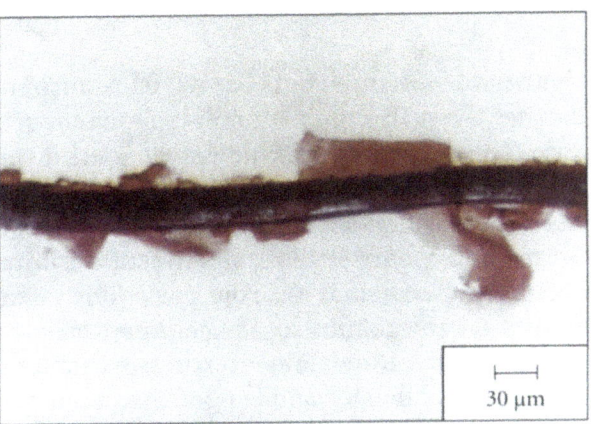

Fig. 84. Coagulated, dyed durable antistatic on a synthetic fiber.

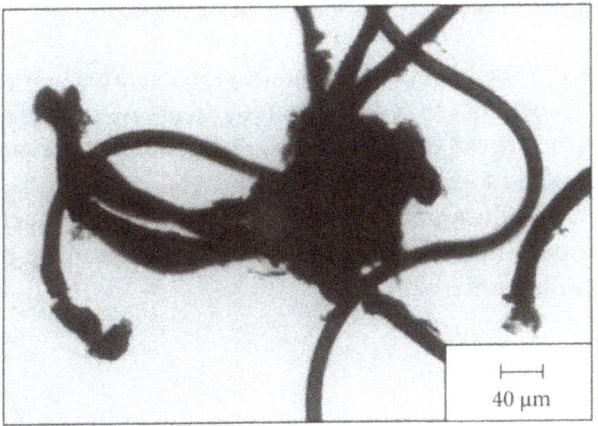

Fig. 85. Polyamide fibers destroyed by acid.

acetic acid is added and the fabric is left in the solution for 10 minutes at room temperature. Acid-damaged polyamide fibers are dyed intensively red by this reagent while thermally damaged fibers remain undyed, Fig. 86.

It was interesting to see that the defect only occurred in black-dyed fabrics. As was discovered later, these were not dyed by adding acetic acid but with formic acid; moreover, the concentrated acid was added to the dyeing liquor in an inappropriate way. If splashes of concentrated formic acid hit the fabric, the polyamide fibers are dissolved. Through dilution of the acid with water, the dissolved polyamide precipitates as an amorphous substance and rebonds the areas.

2.4.2 Detection of Acid-Damaged Polyamide Fibers by Means of Imprints – Practical Example

Fig. 87 – 88
Page 64

After dyeing, light stains with hardened areas could be seen on texturized polyamide fabric. Fig. 87 illustrates the surface imprint on a polystyrene film of a perfect section of the fabric. The deformation of individual fibers during the texturizing process can be clearly recognized. The fibers have partially taken the shape of a corkscrew. This image is typical of polyamide fibers after texturizing by means of the false-twist method. Figure 88 illustrates the surface imprint of a hardened area. It can be recognized clearly that the polyamide fibers have been destroyed and only form an amorphous substance. Apart from this the light areas were dyed red by the test dye Rhodamine B extra (see chapter 2.4.1). Consequently, there is no doubt that the polyamide fibers had been damaged by acid splashes [28].

2.4.3 Acid Damage to Polyamide Stockings – Practical Example

Fig. 89
Page 65

In stockings made of polyamide fibers, acid damage is also relatively frequent. Figure 89 shows a lady's stocking made of polyamide fibers with small holes. The edges of the holes in the brown-dyed stockings were colored white, i. e. they had been decolored. In addition, the polyamide fibers at the edges of the holes were badly deformed. Such deformations, Fig. 89, are not necessarily characteristic of acid-damaged polyamide fibers. Acid damage could only be proven by means of dyeing with Rhodamine B.

2.4.4 Polyester Fabric with Acid Damage – Practical Example

Two pieces of fabric which were dyed orange and red respectively and allegedly consisted of polyester/wool, developed bluish and greenish stains after
finishing, which ran in the direction of the weft, Fig. 90.

Fig. 90
Page 65

A microscopic examination showed that there was no fiber blend. The two
fabric samples consisted of pure polyester fibers. The stains which were
complained of showed a strongly acid reaction, where a pH value of 3.5
was measured potentiometrically. By pressing Congo Red paper onto the
wet fabric, a mineral acid was detected which was identified as sulphuric acid
[29].

Polyester fibers were taken from the spotted areas; under the microscope very
bad damage could be recognized. From some fibers, pieces had virtually been
broken off. Apart from this the fibers were dyed intensively red with m-cre-
sol/Fat Red 5B (Hoechst), Fig. 91.

Fig. 91
Page 65

According to Stratmann, m-cresol/Fat Red is a specific reagent for the detection
of polyester fibers [30]. The examination is carried out as follows: Reagent: 0.5
g of Fat Red 5B (Hoechst) are wetted with a small quantity of methanol and
then dissolved in 50 ml of m-cresol.

Execution of the examination: with blunt scissors, small fiber pieces are squeezed off and treated with the reagent for 5 minutes at room temperature. After
thorough rinsing with methanol, they are briefly rinsed with acetone and once
more washed with methanol.

The fiber ends of the polyester fibers are dyed intensively red by this reagent.
The reaction is particularly successful if the fiber ends are not cut cleanly with
sharp scissors but squeezed off with blunt scissors [Fig. 92]. It should be mentioned in this context that with this test fibers of the type Kodel (Eastman) and
Vestan (Bayer), based on poly-1,4-dihydroxy methylcyclohexaneterephthalate,
are dyed red throughout and not only at the fiber ends.

Fig. 92
Page 67

Model tests showed that it could have been 80–90% sulphuric acid. As
transpired later, the pieces had been carbonized by mistake. Probably after
squeezing of the acidified pieces and their subsequent batching, so-called water
bags had formed in which the acid could concentrate during drying of the
pieces. In the wetter areas of the piece, the acid concentration rose further in
the crabbing oven, thus leading to varying acid-induced attack of the fabrics
which caused the local damage of the fiber material and therefore the stain
formation.

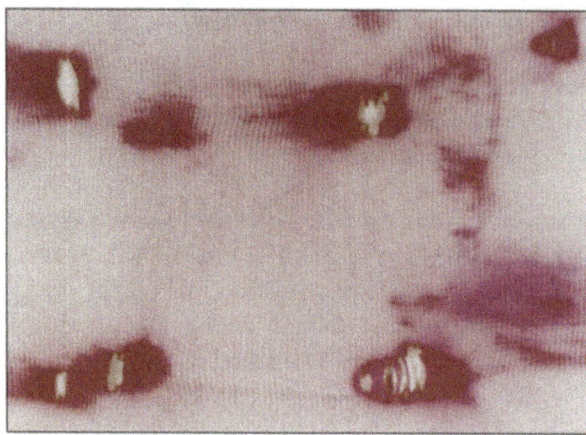

Fig. 86. Acid-damaged polyamide knitwear, dyed with Rhodamine B extra.

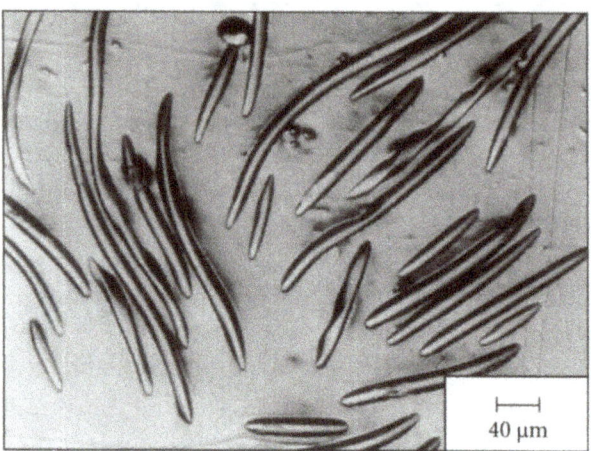

Fig. 87. Film imprint of knitwear made from textured polyamide.

40 µm

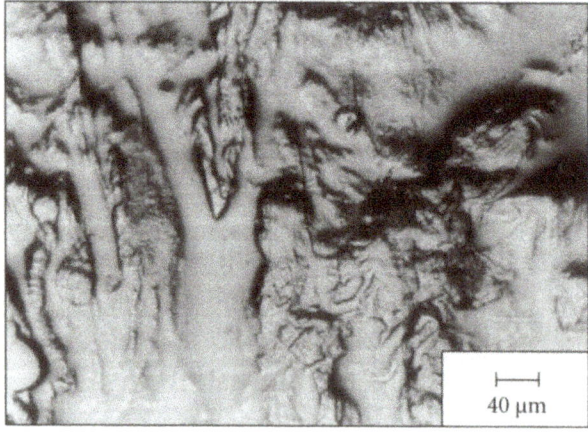

Fig. 88. Film imprint of the fabric from Fig. 87, area with acid damage.

40 µm

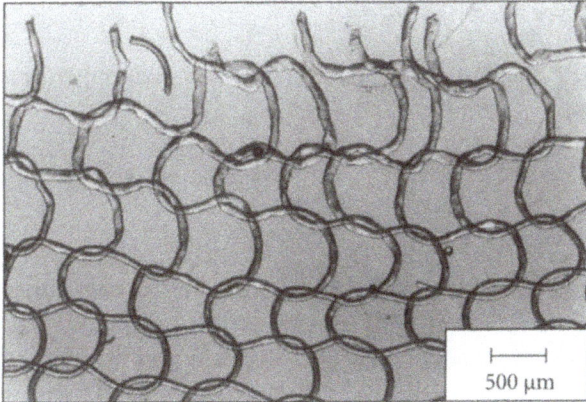

Fig. 89. Outer region of a hole in a polyamide stocking, caused by acid action.

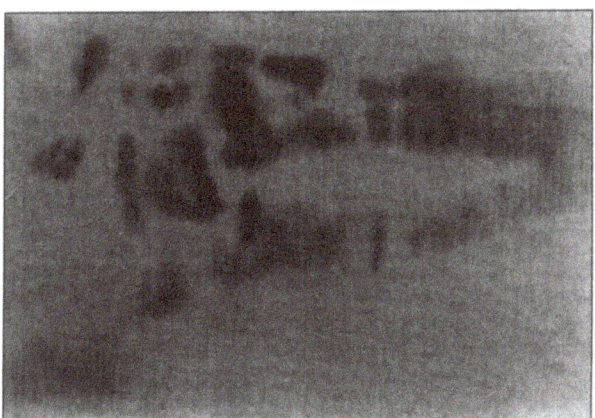

Fig. 90. Polyester fabric with stains after finishing, caused by sulphuric acid.

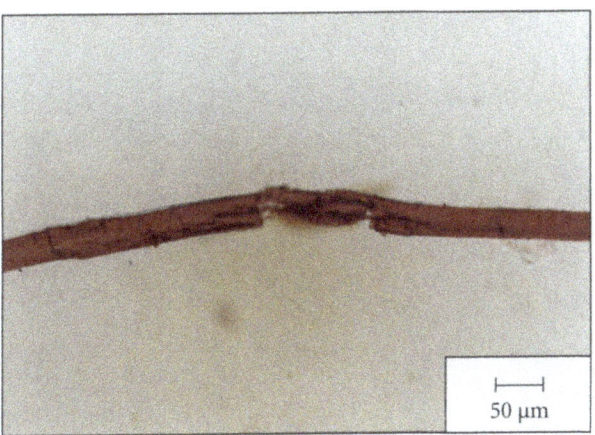

Fig. 91. Acid-damaged polyester fibers from a spotty area of the fabric, Fig. 90. Staining with m-cresol/Fat Red 5B (Hoechst).

2.4.5 Detection of Saponified Acetate Fibers

Acetate filaments are saponified by alkali. One can detect the damage through staining with a substantive dye:

0.5 g/l Sirius Red 4B (Bayer) (C.I. Direct Red 81),
1.0 g/l Glauber salt,
1/2 hour at 50 °C.

Fig. 93
Page 67
Fig. 93 illustrates a fabric made of triacetate which was locally saponified by alkali splashes. The saponified areas were stained red according to the specified method while the unsaponified fibers remained undyed. However, it must be borne in mind during the test on saponified acetate filaments that finishing with the so-called S-finish of triacetate ("saponification finish") deliberately creates surface saponification.

According to Stratmann [2], one can also detect saponified acetate filaments under the microscope by immersing the fibers in zinc chloride-iodine solution. Due to saponification, the outer skin of acetate fibers consists of a regenerate cellulose which is dyed blue by zinc chloride-iodine solution. The acetyl cellulose in the fiber interior dissolves with the reagent and cracks the blue-dyed skin of the regenerated cellulose so that dissolving acetyl cellulose can escape.

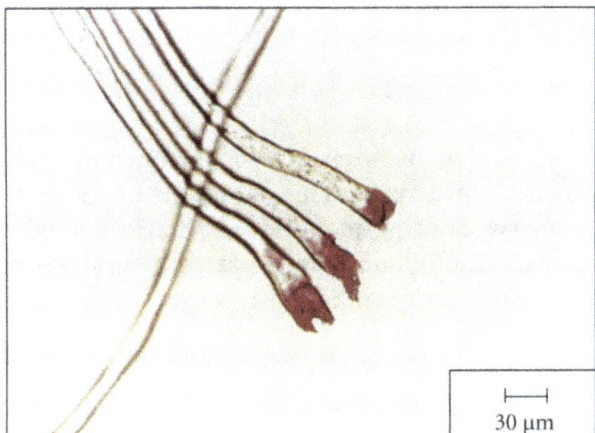

30 μm

Fig. 92. Undamaged polyester fibers after treatment with m-cresol/Fat Red 5B (Hoechst). The staining of fiber ends is typical of polyester fibers.

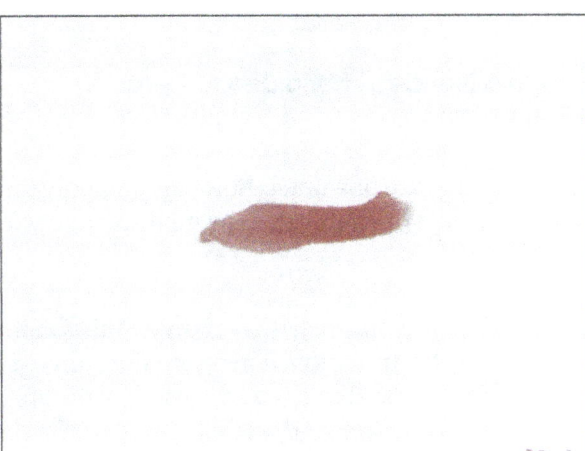

Fig. 93. Acetate filament fabric, locally saponified. Detection through staining with a substantive dye (Sirius Red 4B).

3 Mechanical Damage

During textile finishing, fiber damage is often caused by abrasion [31], resulting e.g. in grayed spots and light streaks. This damage is often only recognized in a later finishing stage when large quantities of material have already been damaged. Natural fibers and cellulose regenerated fibers are more susceptible to mechanical damage than synthetics; however, synthetics also suffer damage.

3.1 Mechanical Damage to Wool

3.1.1 Mechanical Damage to a Blended Fabric Due to Abrasion on the Winch - Practical Example

A wool fabric consisting of a fiber mixture of wool, rabbit hair and polyamide 6 showed unlevelness after dyeing. Apart from this there were thin areas without surface wool.

Fig. 94
Page 69 Microscopic examination of this fabric showed that many wool fibers were split, Fig. 94. There was also fibrillar delamination. Therefore the damage must have been mechanical; it was later discovered that it was caused by the abrasion of the fabric on the winch. Chemical damage to the wool could not be detected. Occasionally, rabbit hair also displayed mechanical damage, whereas the polyamide fibers remained absolutely intact.

3.1.2 Light Streaks and Stains on a Wool Fabric - Practical Example

Fig. 95
Page 69 A fabric sample made of pure wool was examined. It was dyed dark red and showed light streaks and/or stains. A fabric imprint was prepared which revealed that the wool fibers were split at these locations (Fig. 95). In addition, all damaged wool fibers were located at the surface. According to the swelling reaction with ammonium potassium hydroxide, there was no acid damage to the wool. Thus the splitting of individual wool fibers into spindle cells could only have been caused by mechanical action, i.e. abrasion. On split wool fibers incident light was scattered diffusely, thus giving the impression of lighter dyeing.

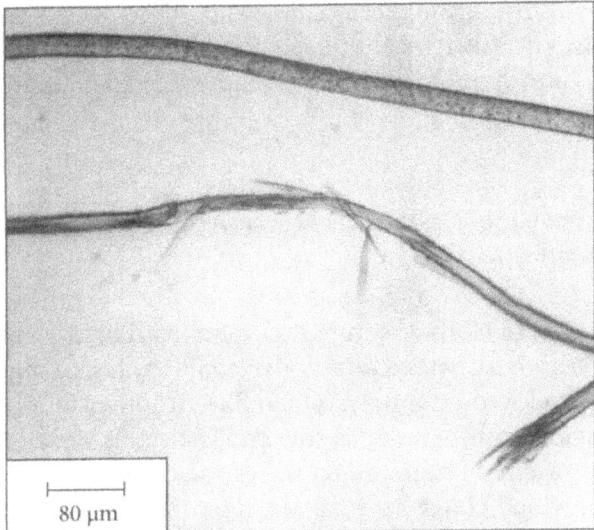

Fig. 94. Wool fiber (below) damaged by abrasion of a blended wool fabric on the winch; above: intact polyamide fiber.

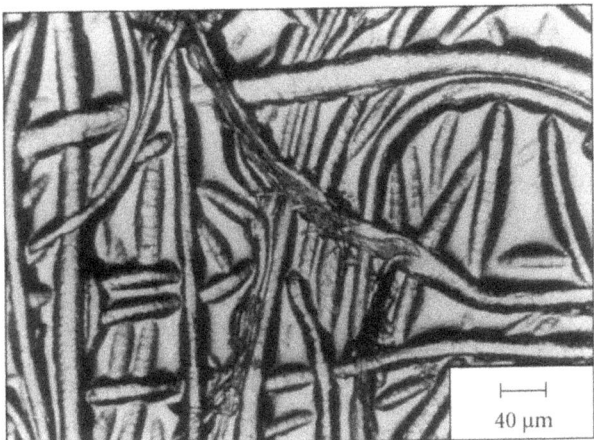

Fig. 95. Film imprint of a wool fabric with mechanically damaged wool fibers.

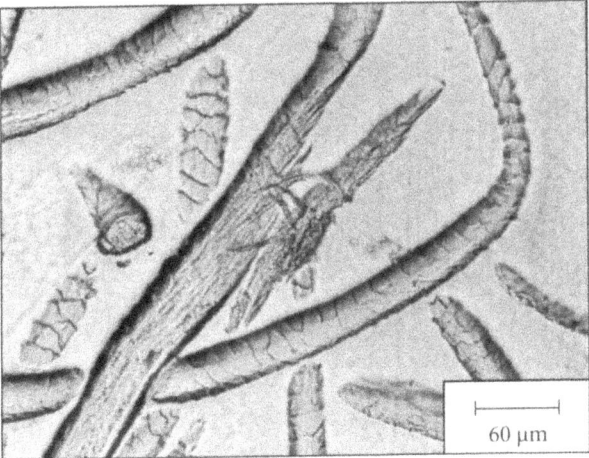

Fig. 96. Film imprint of a wool cloth with light spots due to mechanical damage of individual wool fibers during shearing of the piece.

Fibrillar dissolution of wool into spindle cells can also be observed in bacterially damaged wool. However, it is typical of bacterial damage that only one half is decomposed, while the other remains intact (chapter 9, Fig. 319). In this case there was no such fiber damage.

3.1.3 Light Stains on a Wool Cloth Caused by Mechanical Damage During Shearing – Practical Example

After finishing, a dark blue wool cloth was in part covered with light stains. It was assumed that these areas were reserved during dyeing. First, a large imprint of the stained fabric was produced on a polystyrene film. Another fabric was subjected to grease extraction with petroleum solvent. The stains were found both on the imprint of the original fabric and on the extracted piece. Thus oily and/or greasy substances, which could have caused reserving during dyeing, could not be responsible for the stains. Therefore, the light spots must have been caused by a local structural change of the fabric.

In order to verify this, a stained fabric sample was treated with Pauly reagent. The lighter spots were dyed reddish brown by the dye reagent, thus proving that the original wool was damaged.

Fig. 96
Page 69
For microscopic examination of the fiber material, imprints were produced on polypropylene films because of the better depth of focus. Examination of these imprints confirmed that the wool fibers were clearly damaged in the stained areas. The damage had obviously been caused during shearing of the pieces, because pieces of individual fibers had been cut out and/or chopped off, thus exposing the spindle cells, Fig. 96. This fiber image is typical of mechanical damage caused during shearing of the wool.

3.1.4 Mechanical Damage Caused by Tearing Wool, Comparison with Recovered Wool

Fig. 97
Page 72
During raw wool scouring, individual fibers can be damaged by felting and tearing of the fibers due to deficient transport equipment between the tubs. This damage is not detectable in the finished piece, thus it seldom leads to complaints. Wool fibers can also be mechanically damaged during the spinning process, e.g. on the carding machine. Under the microscope, this shows in the form of brush-shaped ends of the fiber as well as split fibers, Fig. 97.

Recovered wool which is produced by tearing of used textiles (rags) as well as by processing of spinning and weaving mill waste, shows the same symptoms of mechanically damaged wool, i.e. brush-shaped fiber ends, peel-offs and splittings. These products typically contain fibers of different fineness and origin with unequal lengths (short and very short fibers) and different color. Often the color of some of these fibers in the material differs from the yarn color; however, this can also be an intended blend. Fabrics which contain recovered wool wear out fast, leading to bare areas, e.g. at the sleeves or pockets.

3.1.5 Mechanical Damage to Wool Fibers Caused by Wool Pests

Although the action of wool pests on the wool fiber must be considered as biological action of organisms on textiles (see chapter 9), mechanical damage to the fiber predominates. Therefore this topic is also dealt with here.

Damage to wool fibers caused by wool pests such as the larvae of the clothes moth, as well as the fur beetle and the carpet beetle, must be considered as mechanical damage. These larvae feed on keratin-containing fiber material such as wool, bed feathers, furs etc.. Grub damage can be recognized as small, mostly circular holes in the fabric. Under the microscope, the damage caused by clothes moth larvae, which represent the worst textile pests, can be recognized by the crescent-shaped bulges at the sides of the fibers, fig. 98. In most cases there are only bitten-off fiber ends, Fig. 99. However, these fiber images alone do not prove moth damage. For further proof, attention must also be paid to the existence of excrement crumbs, empty pupa quivers, moth residues, caterpillar residues or pupa residues.

Fig. 98-99
Page 72

3.2 Mechanical Damage to Silk

Both mulberry silk and wild silk are very sensitive to mechanical action such as abrasion and rubbing since the fibrils form a fiber composite which can be split easily. This must be considered thoroughly during degumming and dyeing of the silk. Mechanical influences must above all be avoided during dyeing and bleaching of the degummed silk on the winch, since wet degummed silk is particularly susceptible to mechanical damage [32].

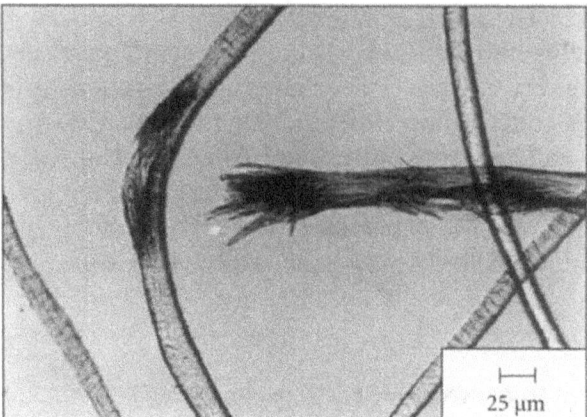

Fig. 97. Mechanically damaged wool fibers. The brush-shaped fiber ends and splittings are especially frequent in recovered wool.

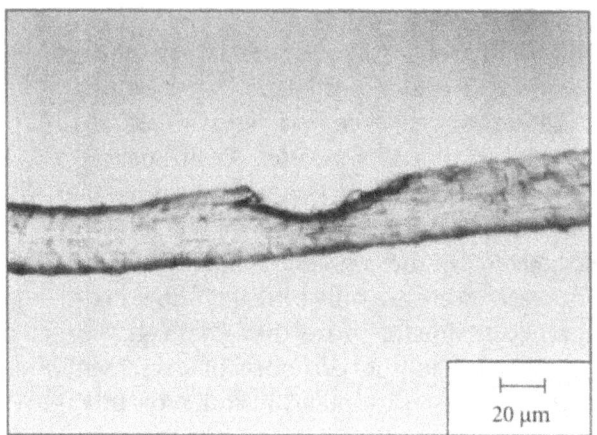

Fig. 98. Wool fiber damaged by the clothes moth larva. The crescent-shaped bite is typical of moth damage.

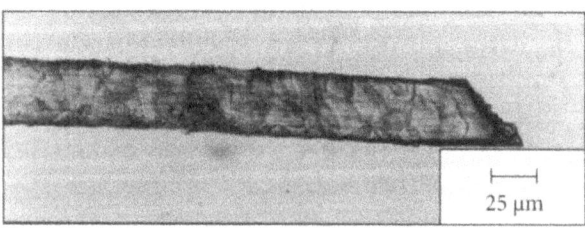

Fig. 99. Wool fiber bitten through by the clothes moth larva.

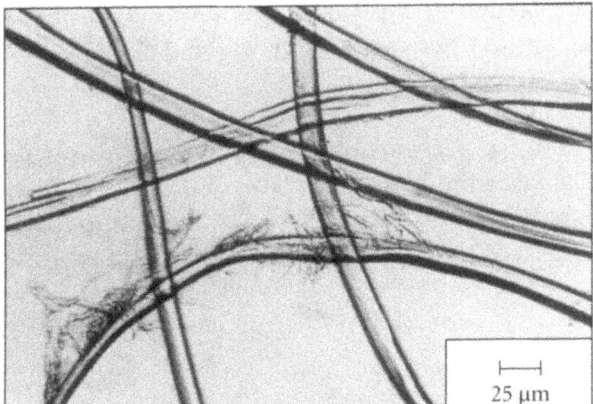

Fig. 100. Mulberry silk, mechanically damaged due to abrasion. Incident light is scattered diffusely on the abraded fiber surfaces. This leads to graying (blanched places).

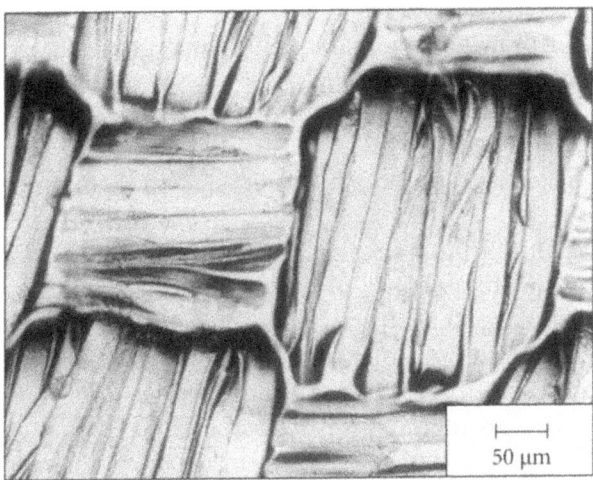

Fig. 101. Film imprint of a non-grayed silk fabric from Tussah. The surface is smooth and clean.

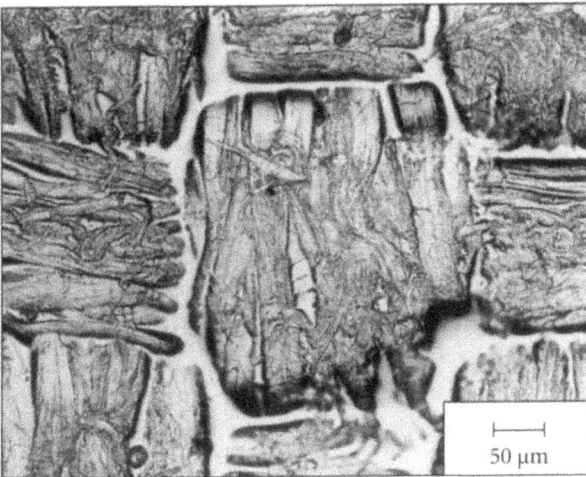

Fig. 102. Film imprint of a silk fabric from Tussah with blanched places. The surface is very rough.

The splits in the fiber structure lead to local graying and/or lighter spots on the silk fabrics; in dark-dyed fabrics there can even be white spots. The graying of silk fabrics is also referred to as blanched places.

Fig. 100–102
Page 73
Under the microscope, splittings, peel-offs and fibrillar dissolution phenomena can be observed in the area of the blanched places, Fig. 100–102. The abraded or split silk fibers scatter incident light diffusely, thus destroying the typical silk luster. This leads to the formation of dull, light or white spots on dyed fabric.

3.3 Mechanical Damage to Cotton

In cotton fibers mechanical damage is more frequent than is generally assumed. Cotton is particularly sensitive to wet abrasion.

3.3.1 Light Streaks on a Black-Dyed Poplin Fabric – Practical Example

Fig. 103–104
Page 75
On a black dyed poplin fabric, light streaks diagonal to the warp direction could be recognized after finishing, Fig. 103. Under microscopic examination it was found that in these areas the cotton fibers showed severe mechanical damage, i.e. abraded fibers and split-offs, Fig. 104. The spliced fibers protruding from the fabric give the impression of a light deposit and/or a lighter coloration of the fabric, since incident light is scattered diffusely at these fiber fragments. According to the positive "pinhead reaction" (chapter 2.3.2, Fig. 72) the fiber material was not chemically damaged. Consequently, splitting must have resulted from abrasion.

3.3.2 Mechanical Damage to a Feather Bed Ticking – Practical Example

Fig. 105
Page 75
A feather bed ticking was rejected because small holes developed after several years of use. It was claimed that splashes of an acid-containing liquid might have stained the feather bed ticking during the manufacturing process. However, the pinhead reaction proved beyond any doubt that the cotton fibers in the damaged areas were not chemically damaged. Apart from this, acid damage to the cotton material during the preparation process could be excluded because this would have shown very soon after production and not after several years. Consequently, the holes cannot possibly have resulted from acid splashes. As shown in fig. 105, the cotton fibers of the feather bed ticking are split and/or

Fig. 103. Black-dyed cotton poplin fabric with a light streak caused by damage to cotton fibers as a result of wet abrasion.

Fig. 104. Abraded and spliced cotton fibers of the black-dyed cotton poplin fabric shown in Fig. 103.

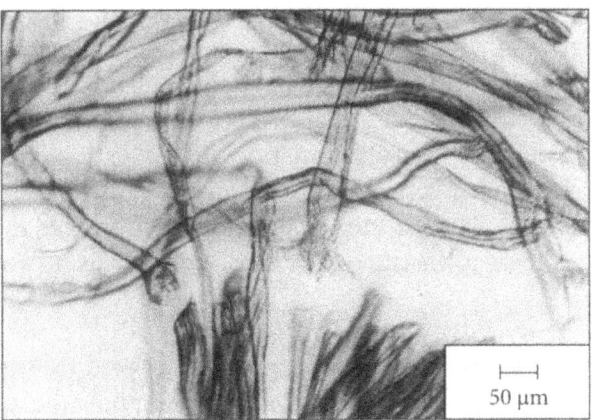

Fig. 105. Cotton fibers, spliced at the fiber ends. They were taken from the outer region of a hole in a feather bed ticking.

fibrillated at the holes. Such damage can only be caused by squeezing and/or pressing. The damage was therefore purely mechanical. Under the low-magnification microscope, Fig. 106, it was found that the broken fibers had been pressed outwards, thus forming craters at the holes, Fig. 106. Pointed objects, probably feather quills, must therefore have been pressed through the fabric from the inside towards the outside.

3.3.3 Graying of a Dyed Cotton Fabric After Extended Use

Fig. 107
Page 77

Graying of heavily worn areas in cotton fabrics (e.g. collars, button facings, collar corners etc.) after extended use and a large number of cleaning processes must be attributed to mechanical damage to the cotton. Figure 107 shows cotton fibers from the sleeves of a decolorized cotton poplin coat with bad mechanical damage caused by abrasion. Incident light is scattered diffusely at the abraded, spliced fibers, thus causing a graying effect.

3.3.4 Mechanically Damaged Cotton Thread of an Oriental Carpet – Practical Example

The following example describes an unusual case of mechanical damage to natural fibers.

Genuine Oriental carpets and rugs had been dry-cleaned and shampooed by a dry cleaning company. Six months after a customer had moved into a new flat, damage in the form of bare patches and clearly lighter pile tufts was observed, especially in those areas subject to particular mechanical strain. The customer presumed that the damage might have resulted from inappropriate treatment during dry-cleaning. The dry cleaner, however, claimed that it might be due to moth damage. The customer's flat had a polyamide velour wall-to-wall carpet with a short resilient pile and underfloor heating. The Oriental carpet was spread on the polyamide velour fabric. Fibers were carefully taken from those areas of the Oriental carpets where the color of the pile tufts had changed visibly. Careful examination of the extracted fibers showed that the decolorized fibers were almost exclusively cotton fibers from the warp of the Oriental carpets. These cotton fibers must have entered the pile after the warp had been damaged in the worn areas, possibly as a result of beating or vacuuming.

Fig. 108
Page 77

While the wool fibers extracted from the damaged areas showed no signs of mechanical or chemical damage (apart from different lengths), the cotton

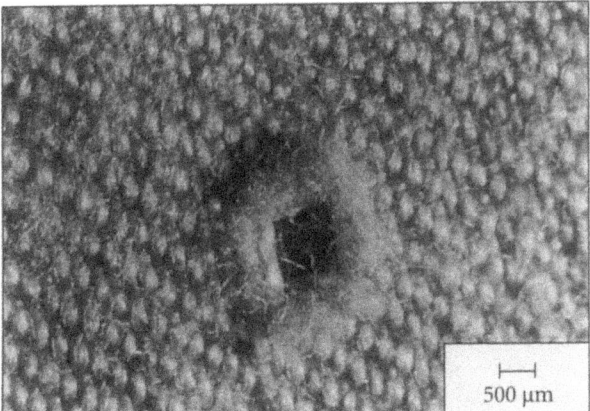

Fig. 106. Hole in a feather bed ticking. A feather quill was pressed through the fabric from the inside towards the outside, thus forming a crater at the outer region of the hole.

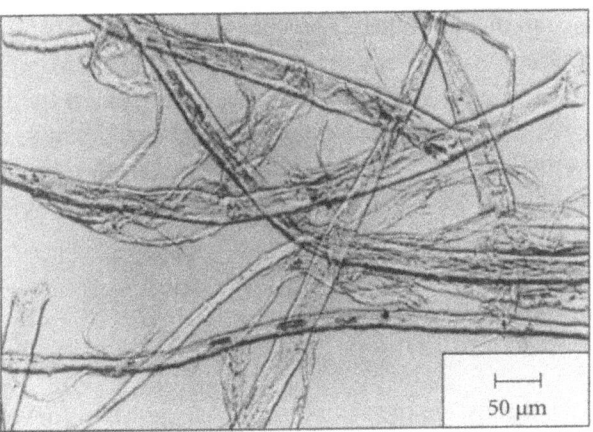

Fig. 107. Cotton fibers from the sleeve of a heavily worn cotton poplin coat, badly damaged due to abrasion.

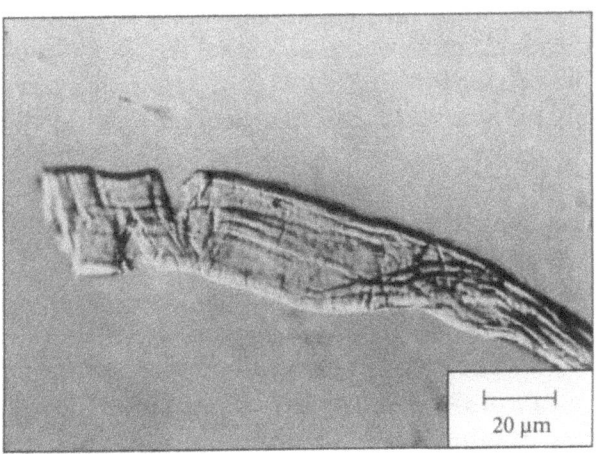

Fig. 108. Cotton fiber from the warp of a mechanically damaged Oriental carpet with notches and splits.

contained a relatively large number of mechanically damaged fibers, Fig. 108. This could clearly be recognized from the notches, splits and incisions in these fibers. Chemical damage could not be detected. The damage was caused by the high strength of the polyamide fiber. The resilient polyamide pile had virtually cut through the cotton warp under mechanical strain, i.e. when walking on the carpets. After the wool carpets were separated from the polyamide pile by means of a protective layer, no further damage occurred [33].

3.3.5 Darker Colored Streaks Due to Squashed Cotton Fibers – Practical Example

Fig. 109
Page 79

After dyeing and drying, some patches of discontinuously pretreated cotton fabrics were dyed darker than the rest of the material. The more intensive color areas were clearly marked. Chemical damage to the fiber material could not be detected. There were no sizing residues. The unlevelness could not be eliminated by removing the dyestuff and redyeing the fabric. Microscopic examination showed that the fibers were mechanically damaged in the respective areas by squashing, Fig. 109. It has repeatedly been found that squashed cotton is dyed more intensively than the undamaged material, while cotton damaged by abrasion appears to be dyed lighter due to optical effects [34] (see chapter 3.3.1, Fig. 104).

3.3.6 Crease Marks

Fig. 110
Page 79

Crease marks represent a special case of mechanical damage which develops in the creases of a swollen, stored fabric. They usually occur in heavy, tightly woven cotton fabrics in the steamer of continuous pretreatment plants which are equipped with a plaiting-down system. In the subsequent continuous dyeing process, crease marks, also referred to as "crows' feet", are recognizable as dark dyed, irregular, line-shaped marks, Fig. 110. Microscopic examination methods which permit recognition of this damage are described in the following.

Fig. 111–113
Page 79/81

Large imprints of fabrics with crease marks showed that local structural changes of the fabric, such as displacement of the weft yarns against the warp threads, as well as flattened threads or deposits of any kind, cannot be responsible for the formation of crease marks. This can be concluded from the fact that the dark dyed line-shaped crease marks are not visible on the imprints, Fig. 111. However, this does not exclude chemical or mechanical fiber damage in the micrometer range. Examination with the aid of a pinhead reaction,

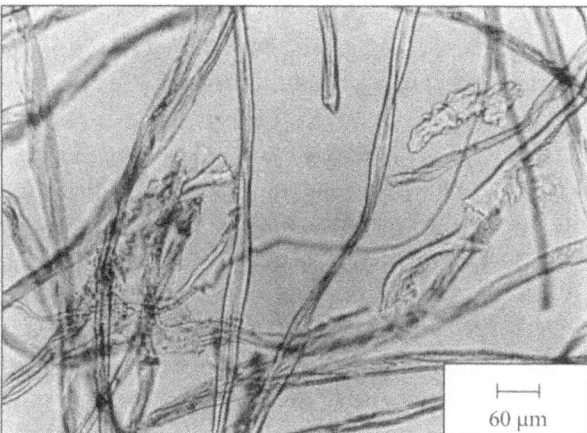

Fig. 109. Squashed cotton fibers, which, due to their changed structure, are dyed darker than intact fibers.

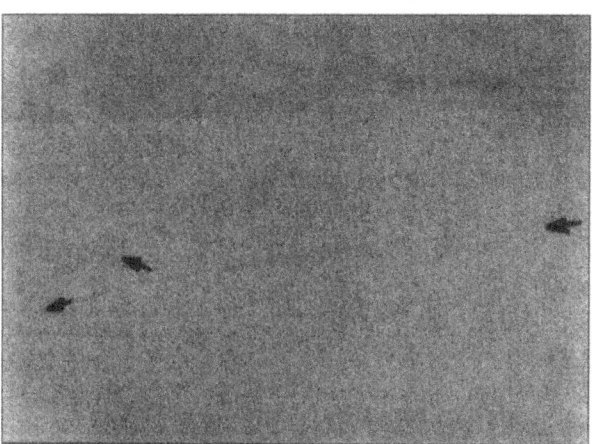

Fig. 110. Crease marks on a cotton twill fabric. Darker dyed, irregular line-shaped marks can be clearly recognized.

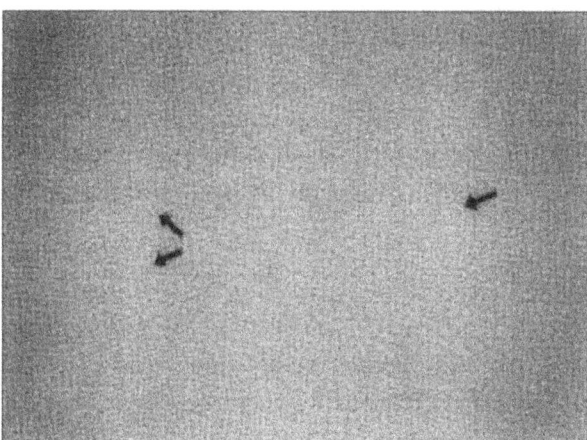

Fig. 111. Large film imprint of the cotton twill fabric from Fig.110. The crease marks (see arrows) are not detectable.

chapter 2.3.2, Fig. 72, showed that chemical damage to the fiber material cannot be responsible for the formation of crease marks. Fig. 112 illustrates the fabric cross-section at a crease mark of a rejected fabric. There are no signs of mechanical damage due to abrasion. It is simply seen that individual weft yarns are slightly flattened and not circular. One might assume that the fabric in these areas was pressed and/or squashed slightly more than in the other areas. However, this is insignificant for the formation of crease marks. Slight thread deformations were also observed on the rest of the fabric. This was especially true with the more voluminous yarn areas which were compressed slightly more by the squeezing rollers. Fig. 113 (fabric cross-section with slightly smaller magnification) shows that the fabric is flat at the crease marks. There are no raised or depressed areas which can normally be found in the case of genuine creases.

Mechanical damage resulting from wet abrasion of the fabric can be easily recognized under the microscope. It shows in the form of split-offs, see chapter 3.3.1, Fig. 104, which lead to optical lightening effects and give the impression of a lighter coating and/or a lighter dye shade. The crease marks, however, are always dark colored streaks, in which, so far, abraded cotton fibers have never been found. Abrasion damage can therefore be excluded as a cause for their development. Since cotton fibers with squashing damage are dyed dark, see chapter 3.3.5, Fig. 109, the obvious conclusion is that the crease marks must have been caused by squashed spots.

Fig. 114–115
Page 81/82

In order to confirm this, a large number of cotton fibers from crease marks and fibers from faultless areas of a fabric were examined microscopically. In many cases the light microscope detected individual cotton fibers with squashing damage in the undamaged fabric, Fig. 114. The imprints of the crease marks also showed cotton fibers with squashed spots, Fig. 115. A detailed examination, however, shows occasional cracks perpendicular to the direction of the thread on the imprints of the crease marks, which hints at surface damage. However, under the light microscope the damage proved to be so small that it could hardly be recognized on the individual fibers. They were only clearly visible on a few fibers. Therefore it cannot be concluded that crease marks are mechanically damaged cotton fibers although their deeper dyeing hints at squashing damage [35].

Fig. 116–120
Page 82/84

With the aid of the scanning electron microscope, the damage to the cotton fibers could be clearly recognized. Figure 116 and 117 display cotton fibers from the faultless parts of the piece. There are only small cracks and/or splits. Squeezing damage is not detectable. With the cotton fibers from the area of the crease marks, however, squashing damage is easily recognizable in addition to

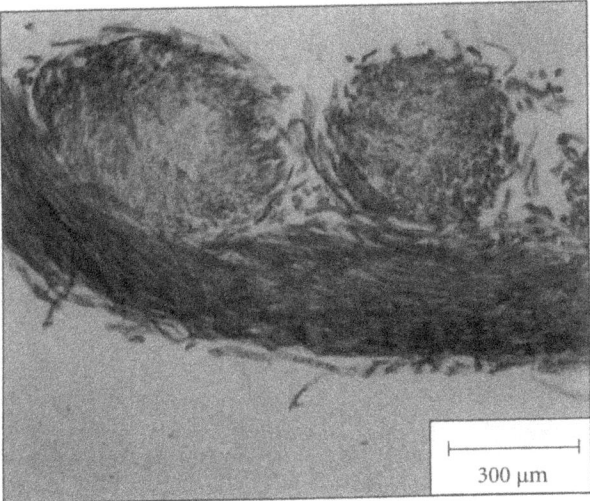

Fig. 112. Fabric cross-section at crease marks of the fabric from Fig. 110. Significant yarn deformations are not detectable.

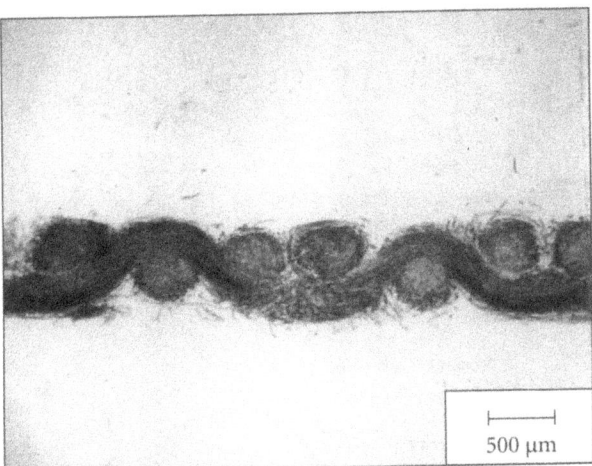

Fig. 113. Fabric cross-section from Fig. 112, smaller magnification. The fabric is flat and has no raised or depressed areas.

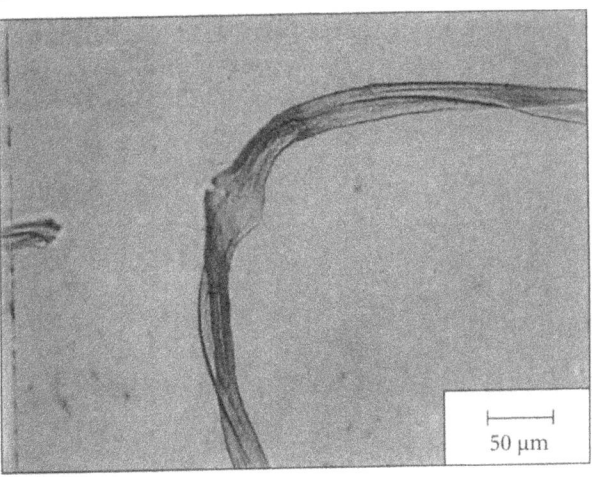

Fig. 114. Fiber preparation from crease marks with locally squashed cotton fiber.

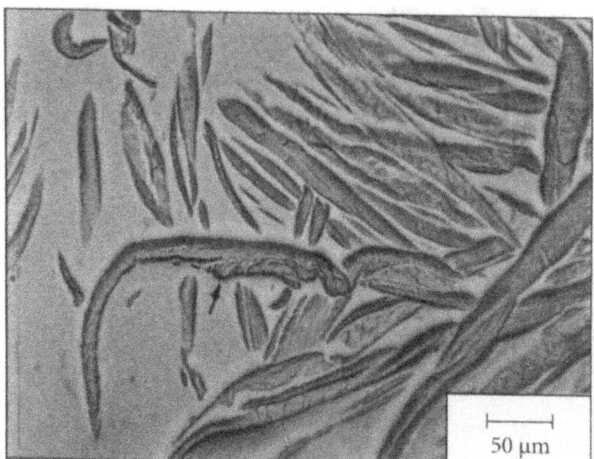

Fig. 115. Surface imprint of crease marks with squashed cotton fiber (arrow).

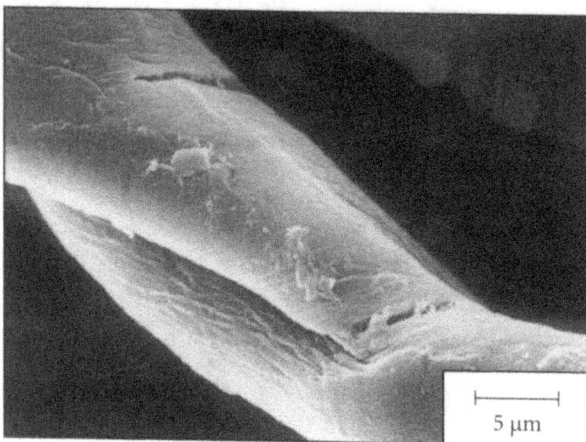

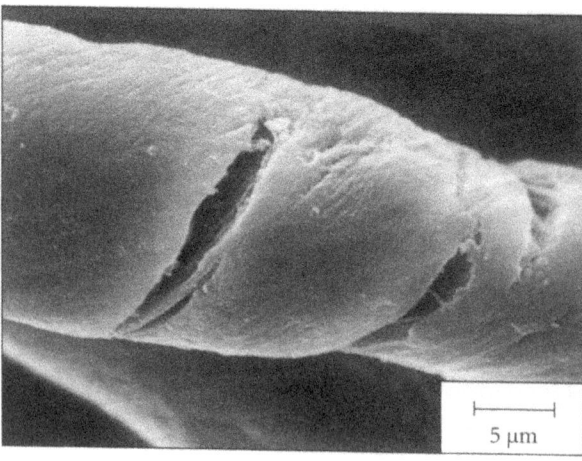

Figs. 116, 117. Micrographs of cotton fibers (scanning electron microscope) with small cracks and splits from evenly dyed areas of one piece.

the presence of notches, cracks and splits, Fig. 118 and 119. This squashing damage is due to the fact that fibers were bent in the areas of the folds during pretreatment in the plaiting-down process. In the compressed regions, strong swelling led to subsequent squashing. The cracks developed during deswelling and stretching of the fibers in the wash boxes through tight strand guidance (Fig. 120).

As a result, by using special products during alkaline impregnation of the fabric it was tried to improve the elasticity and the suppleness of the fibers and thus the resistance to squashing and compressing. This improved the appearance of heavy tightly woven cotton fabrics; however, complete elimination of the crease mark was still not possible. Apparently this problem can only be solved in the machines, for example by using a tight strand guidance in the steamer.

3.4 Mechanical Damage to Cellulose Regenerated Fibers Graying During Dyeing on the Winch – Practical Example

Cellulose regenerated fibers must not be exposed to strong mechanical influences. The effects of mechanical strain are explained by means of an example.

Cellulose regenerated fibers are very sensitive to abrasion, Fig. 121. A viscose staple fiber from a blue dyed viscose staple fabric showed strong graying after dyeing on the winch. The explanation is based on optical effects. The surfaces of the viscose staple fibers were roughened by abrasion of the pieces on the winch. Incident light is scattered diffusely by the surface damage, thus leading to the graying effect. Of course, by removal of dyestuff and redyeing the fabric such damage cannot be eliminated; indeed, it is increased.

Fig. 121
Page 86

3.5 Mechanical Damage to Synthetic Fibers

Mechanical damage to synthetic fibers is rare because they are usually more resistant to mechanical influences than natural fibers and cellulose regenerated fibers. In most cases it is a thermo-mechanical deterioration which is described in chapter 4.

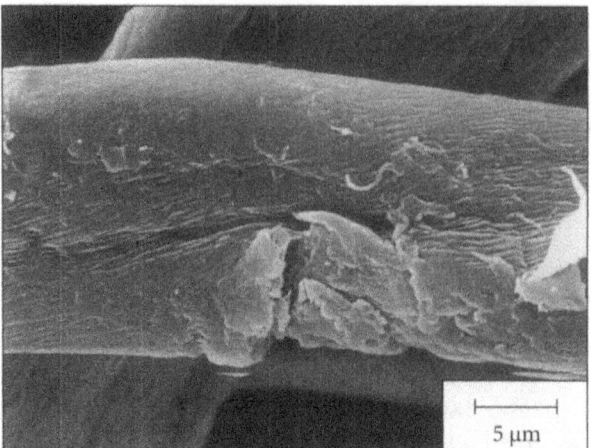

Figs. 118, 119. Micrograph (scanning electron microscope) of cotton fibers with squashing damage from a crease mark.

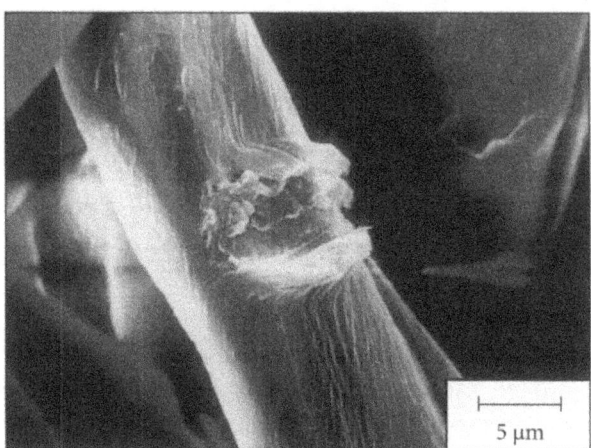

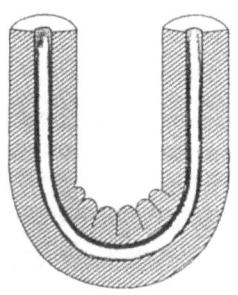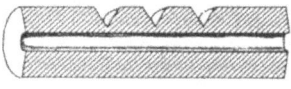

Fig. 120. Schematical representation of a damaged fiber from a crease mark. Left: Compressed and Aquashed regions in the bends of the swollen fiber. Right: Notches and splits after shrinking and stretching of the fiber.

3.5.1 Graying of a Carpet Made of Acrylic Fibers – Practical Example

From use of a wheelchair, different areas of two carpets, one 100% acrylic and the other 42% acrylic/58% polyamide, were subjected to particular strain. Compared to the less worn areas they were clearly lighter and turned into a lighter gray color. At first this was explained by insufficient dyeing of the carpet fibers (16.5 decitex) or by penetration of filling substances from the back coating into the pile of the carpets. Microscopic examination showed that dull-spun and lustrous acrylic fibers had been incorporated in the carpets. In the heavily worn areas the acrylic fibers were badly damaged, which resulted in peel-offs and split fibers, Fig. 122. Incident light was scattered diffusely at the ends of these fibers, thus causing graying and/or lightening of the carpet color. In the case of the carpet with the polyamide fiber blend, only the acrylic fibers displayed damage which was even stronger than that of the 100% acrylic carpet. Under heavy mechanical strain the polyamide fibers had apparently additionally damaged the softer acrylic fibers. The presumption that the filling substances of the back coating had penetrated the surface of the carpets could not be confirmed in the examination.

Fig. 122
Page 86

3.5.2 Graying of a Napped Blanket Material Made of Acrylic Fibers – Practical Example

An acrylic blanket material consisting of a cotton warp and a weft of acrylic fibers could not be napped uniformly. Furthermore, the pieces which were more difficult to nap showed slight graying after napping. The case was also microscopically examined. This showed that some acrylic fibers were badly damaged, the surfaces were abraded, Fig. 123. The presumption that the material showed differences in surface smoothness due to an unsuitable application of softener was confirmed by changing the softening agent.

Fig. 123
Page 86

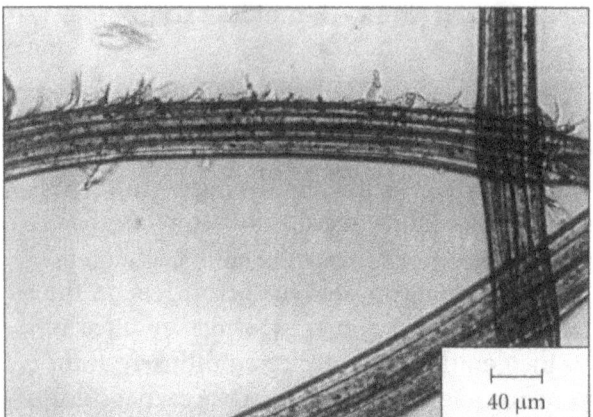

Fig. 121. Viscose staple fibers damaged during dyeing due to abrasion on the winch. The abraded fibers scatter incident light diffusely, leading to graying.

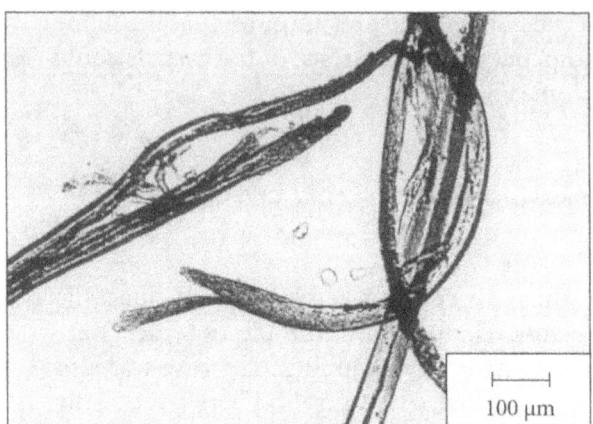

Fig. 122. Spliced acrylic carpet fibers from a heavily worn area of a carpet.

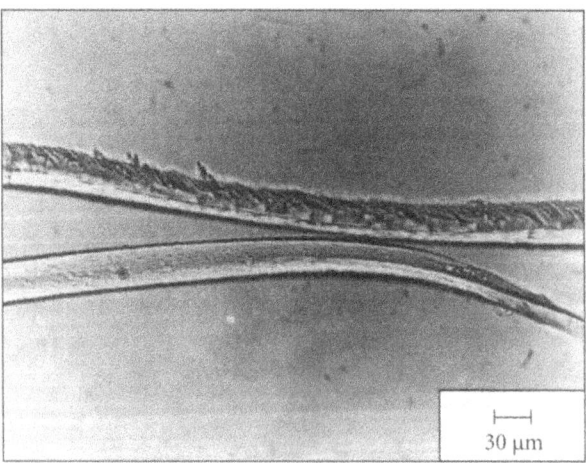

Fig. 123. Acrylic fiber, abraded at the surface during napping due to inappropriate softening. This surface damage caused graying.

4 Thermal and Thermo-Mechanical Damage to Synthetics

A typical characteristic of synthetic fibers is their thermoplasticity. In order to provide the material with optimal use characteristics, heat treatment in the thermoplastic range cannot be avoided during manufacturing, finishing and garment production. Irreversible damage can only be prevented if the softening range of the fiber polymers is considered thoroughly. In addition, synthetics can suffer irreversible deformation during all processes which generate friction heat. Pressure or impact, causing a temperature increase, can also lead to the deformation of synthetics [36, 37].

4.1 Thermal Damage Caused by Direct Heat

Thermal damage to synthetics can be caused during setting, drying, texturizing, singeing and ironing.

4.1.1 Thermal Deformation of Synthetic Fibers During Setting

The setting of synthetic fibers by means of heat is an important and necessary finishing process. Through setting, synthetics are stabilized in their shape so that they do not crease and shrink under usual washing conditions. Since setting is one of the processing steps which take place in the thermoplastic range of the fibers, thermally deformed fibers are particularly frequent in those areas in which the fibers in the yarn cross, Fig. 124. These slight local deformations have no effect on the color evenness or the general appearance within one piece, as long as the setting temperature and time are kept constant. The following practical example illustrates problems that can result if these requirements are not fulfilled.

Fig. 124
Page 88

4.1.2 Weft Streaks in a Fabric Made of Polyester/Wool After Setting
– Practical Example

A fabric made of polyester/wool was rejected because of dark weft streaks. In the blended fabric only the wool portion was dyed pink. Preliminary examin-

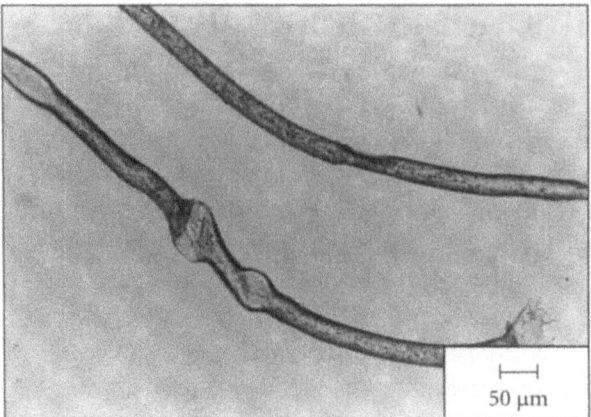

Fig. 124. Thermal deformation of the polyester fibers in crossing areas in the yarn.

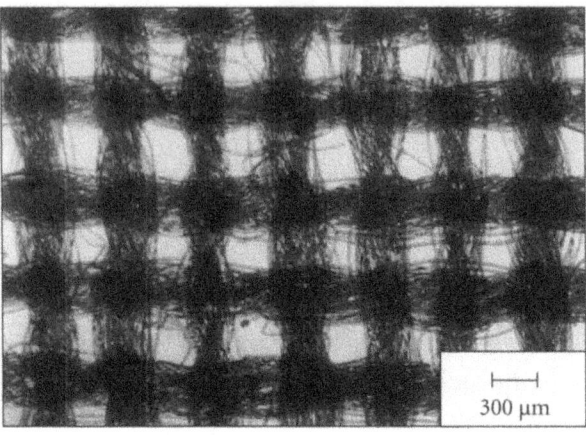

Fig. 125. Fabric made of polyester/wool after extraction of the wool portion with chlorine lye. Some polyester fibers display melted spots caused during singeing of the fabric.

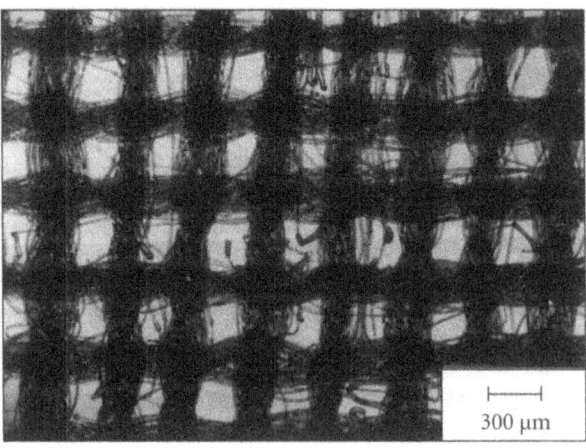

Fig. 126. The same fabric as in Fig. 125. Micrograph of a streaky, grayed area with polyester fibers which were thermally deformed and melted by overheating during long rest periods on the tenter frame during setting.

Fig. 125-126
Page 88

ation of the fabric showed that in some places the weft streaks were not always parallel. Microscopic examination of the fiber material showed that the wool portion in the fabric was intact. After the wool portion had been dissolved using chlorine lye, the same weft streaks were still recognizable on the remaining transparent polyester fabric. Under the magnifying glass microscope in transmitted light, in the intact areas only isolated melt balls on the polyester fibers could be observed, Fig. 125. The reason for this was that the fabric had been singed previously (see chapter 4.1.6). At the weft bars, however, numerous polyester fibers were deformed, Fig. 126. Thus the fabric must have been locally overheated. A check in the plant revealed that the nozzle drying system had not been switched off during longer rest periods of the fabric on the tenter frame.

4.1.3 Graying in a Woven Fur Made of Polyvinyl Chloride and Acrylic Fibers After Drying – Practical Example

Polyvinyl chloride fibers are extremely temperature-sensitive. On account of their shrinking power, they are, among other things, used for the production of woven furs. The temperatures at which the fibers can be finished are relatively low. An excessive drying temperature during finishing can lead to serious defects. A practical example will prove this:

Fig. 127-128
Page 90

A woven fur consisting of 50% polyvinyl chloride fiber and 50% acrylic fiber was rejected because of graying of the black dyed stripes in the fur. At first this defect was attributed to insufficient rinsing of the sizing agent, but then it turned out that polyvinyl alcohol had been used as a sizing agent. Since these sizing agents are usually easy to wash out, it was quite improbable that the defect had been caused by the sizing agent, particularly since only individual pieces of the woven fur showed the irregularity. A drying temperature of 80 °C was prescribed for the fur material. Model tests proved that the graying of the fiber material grew stronger with an increasing excess in drying temperature; the PVC fiber material also shrank to an ever increasing extent. Microscopic examination of the rejected fabric and the original sample proved that the pile of the damaged fabric had clearly shrunk as compared to the original sample [38]. This could be recognized clearly on the isolated pile fibers, Fig. 127, and the cross-sections, Fig. 128. The damage was thus caused by temperature irregularities during drying, which explains why the graying only occurred in individual pieces and/or lots of the material.

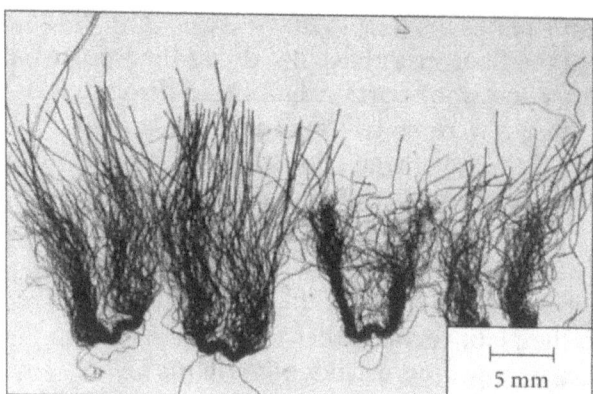

Fig. 127. Pile fibers from a black dyed woven fur consisting of PVC and PAC fibers. Due to temperature variations beyond the permissible maximum during drying of the pieces, the pile fibers shrank variably. Left: Pile fibers from the original sample. Right: Pile fibers from the grayed areas.

Fig. 128. Fabric cross-sections of the woven fur in Fig. 127. Left: Significantly shrunk pile from a grayed area after drying. Right: Original sample.

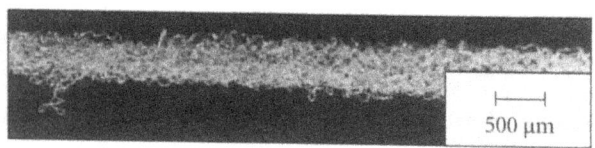

Fig. 129. Textured polyamide yarn; a voluminous yarn results due to crimping.

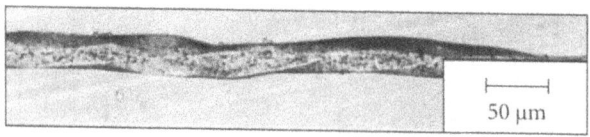

Fig. 130. Polyamide fiber consisting of a textured yarn with deformations similar to those that result during false-twist texturizing.

4.1.4 Thermal Deformations During Texturizing

For texturizing, the thermoplasticity of synthetics is exploited. Deformations of the fiber material are therefore unavoidable. This is explained using the false twist method as an example. During the process, the yarn is strongly over-twisted at a temperature in the thermoplastic range. The deformable fibers which are soft in the heating zone change into the spiral form without tension. After cooling, the fibers are downtwisted, the deformation being set. A voluminous yarn results, Fig. 129. In addition to the deformation of the individual fibers into spirals, the fiber cross-section also changes. Under the microscope it can be seen that a round fiber has changed into a polygonal shape. One can also recognize this change from the side, although it is less obvious, Fig. 130.

Fig. 129–130
Page 90

Figure 131 displays the cross-section of a textured polyester yarn. The black and white photo of the angular fiber cross-section shows that the badly deformed areas are dyed darker than the less strongly deformed areas. In this way temperature variations during texturizing can be recognized. In spite of careful production monitoring they cannot always be completely avoided. Even if they are small and not detectable at the fiber cross-section, they influence the physical characteristics of individual threads. In the case of fabrics and knitwear made of textured yarn the variations lead to streaky dyeing.

Fig. 131
Page 92

4.1.5 Light Stains on a Printed Cotton Fabric Due to Melted, Flat Rolled Polypropylene Fibers – Practical Example

In a printed cotton fabric, undyed white fiber particles were detected. In order to identify these white fiber particles with a microscope, a zinc chloride-iodine solution was used which dyes cotton fibers blue. The rejected fiber particles were not dyed by the reagent, Fig. 132. Apparently they are melted and fused synthetic fibers, rolled flat to form a film, which are insoluble in Cuoxam, glacial acetic acid (cold and boiling), 6N hydrochloric acid, concentrated nitric acid and concentrated sulphuric acid. The melting point of these fibers was 165 °C, which is characteristic of polypropylene fibers. Obviously, the small number of polypropylene fibers contained in the warp and weft yarn were impurities from the spinning mill and had not been added deliberately. Although only few of these fibers were contained in the warp and weft yarn, they caused the unsatisfactory appearance of the finished product.

Fig. 132
Page 92

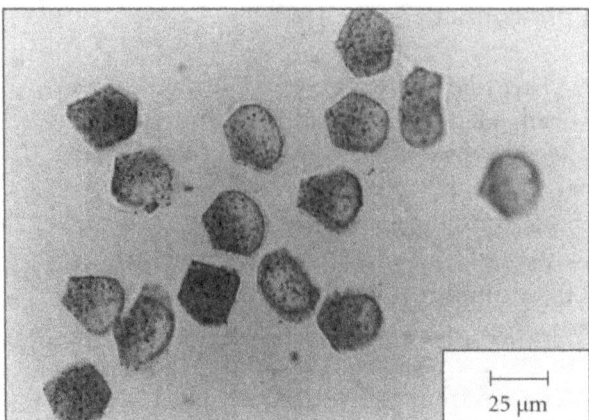

Fig. 131. Cross-section of a dyed, textured polyester yarn. The cross-section of the originally round fibers is angular; the badly deformed fibers are dyed darker.

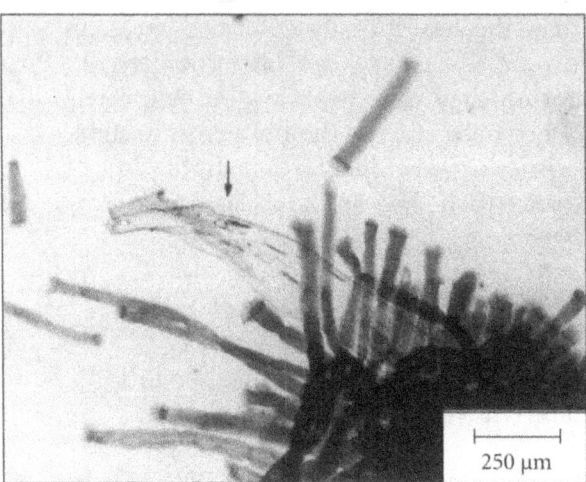

Fig. 132. Cotton fibers, immersed in zinc chloride iodine; the thermally deformed, flat-rolled polypropylene fiber (arrow) is not dyed by the reagent.

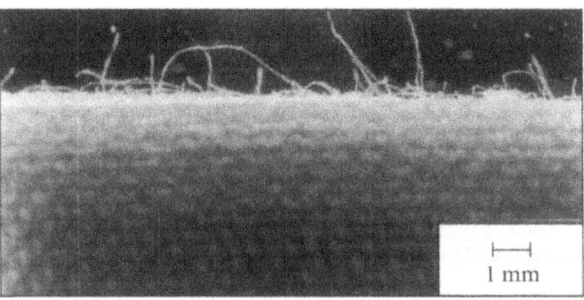

Fig. 133. Unsinged feather bed ticking made of pure cotton; the crease of a folded fabric sheet under the stereo microscope; the surface of the textile fabric is hairy.

4.1.6 Singeing Damage to Synthetic Fibers

In addition to shearing, singeing is a well-proven method for the production of smooth and clean fabric surfaces which are free of fluff. Fabrics from the loom have a more or less hairy and/or fluffy surface, Fig. 133. To remove fiber fluff, the fabrics are singed by drawing them through a non-luminous gas flame. The fibers protruding from the fabric are burnt off, Fig. 134, so that a smooth fabric surface results. Fig. 133–134 Page 92/94

With blended fabrics made of polyester/cotton, polyester/rayon staple, polyamide/cotton or polyester/wool, which are also singed in order to reduce their pilling, one must take into consideration that synthetics, in contrast to cellulosic fibers and wool, do not burn but melt. Spherical fibers are formed, whose dyeing behavior is completely different from that of unsinged fibers, Fig. 135. In the case of printed fabric made of polyester/wool, for example, the fabric is often cropped in order to remove the melt balls. Furthermore, it must be considered that immediately after the singeing process the synthetics are still soft and deform easily. If after singeing the fibers have not cooled down properly, and run over rollers under tension (guide rollers) and pressure ("spark killer squeezing roller") (e. g. through excessive feed rates) in the worst case they can be squeezed to form a film. Fig. 135 Page 94

An uneven moisture content of the fibers within a piece can also lead to an uneven singeing effect since moist fibers are not burnt at all, or only incompletely under the conditions of the singeing process. In this case the resulting fluff has an uneven thickness, thus causing optical effects due to differences in light scattering. In addition, fibers protruding from the yarn and/or fabric are often dyed darker, leading to an inferior fabric appearance. Singeing damage can also result from an uneven flame height and variations in manufacturing speed. Weaving defects such as raised and depressed areas in the textile fabric can cause additional difficulties during singeing [39].

4.1.6.1 Dark Stains on a Gabardine Fabric After Singeing and Dyeing – Practical Example

After dyeing and finishing a gabardine fabric made of polyester/cotton showed dark stains, irregularly distributed throughout the fabric. Microscopic examination showed no encrustations or deposits but revealed thermal deformation and melt balls on numerous polyester fibers. The thermally deformed fibers and hardened areas of the melted fiber ends were clearly dyed darker; this is also shown on the black and white photos, Fig. 136. The polyester Fig. 136 Page 94

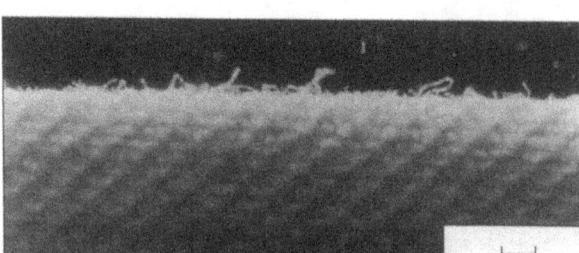

Fig. 134. Singed feather bed ticking as shown in Fig. 133; the surface is smoothed by singeing of the protruding fibers.

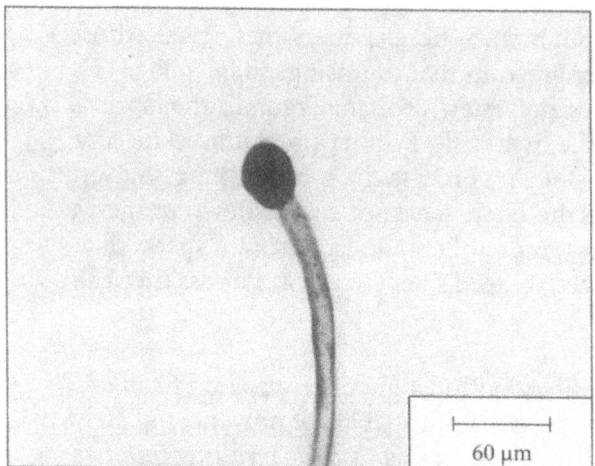

Fig. 135. Polyester fiber from a singed fabric made of polyester/cotton with darker dyed melt balls.

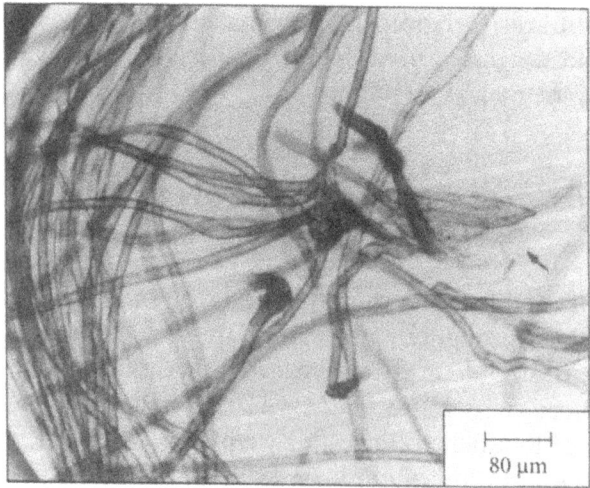

Fig. 136. Fiber preparation of a singed polyester/cotton fabric with stains after dyeing, with melted polyester fiber ends that are dyed darker and a polyester fiber rolled into a transparent film (arrow).

fibers were occasionally rolled into a thin skin. Thus this fabric was not damaged during dyeing but during singeing.

4.1.6.2 Stain Formation and Film-Like Coating Due to Thermally Deformed, Flat-Rolled Polyester Fibers – Practical Example

A fabric made of polyester/cotton was covered with dark stains after singeing and dyeing. In these areas, film-like coatings could be mechanically removed.

Microscopic examination showed that the fiber material in the spotted areas was strongly encrusted and bonded with a film-like substance. With m-cresol/Fat Red these deposits were clearly dyed red, Fig. 137, i.e. not only the fiber ends squeezed off with blunt scissors, see chapter 2.4.4, Fig. 92. It was found that the melting point of the film-like deposits lies between 255 °C and 260 °C. Consequently it was highly probable that the deposits were a melted polyester fiber substance. Figure 138 displays the film imprint of one of the spotted areas. On the imprint it can be clearly seen that it is a melted polyester fiber substance, partially rolled into a film. The textile defect had been caused during singeing of the fabric.

Fig. 137–138
Page 96

4.1.6.3 Streaks and Stains on a Singed Fabric Made of Polyester/Viscose Staple – Practical Example

Piece goods made of polyester/viscose staple dyed by means of the thermosol process, showed small gray stains and streaks after padding. It was first determined whether the defect was located on the viscose staple or on the polyester. For this purpose the viscose staple portion of a sample was removed by means of the formic acid/zinc chloride process [40]; stains and streaks were even more clearly visible than before.

Microscopic examination showed that the gray stains and streaks could apparently be attributed to the formation of numerous melt balls from the polyester fibers. Figure 139 shows an imprint of the polyester/viscose staple fabric. The polyester fibers can be distinguished easily from the viscose staple fibers with the typical longitudinal grooves. A melt ball on the end of a polyester fiber can be seen. This proves that the fabric was singed before dyeing.

Fig. 139
Page 97

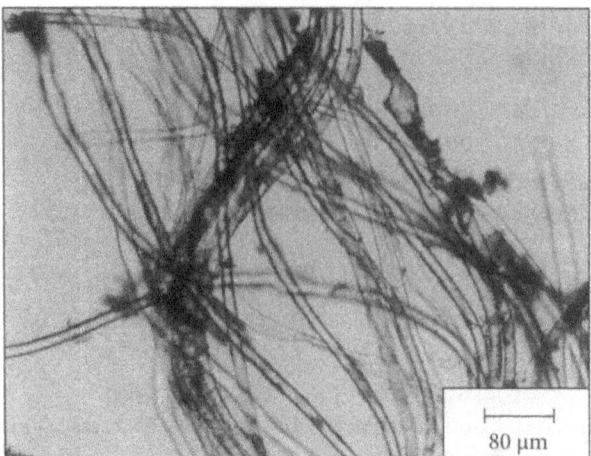

Fig. 137. Fiber preparation made of a polyester/cotton fabric with thermally deformed, partially film-like flat-rolled polyester fibers dyed with m-cresol/Fat Red.

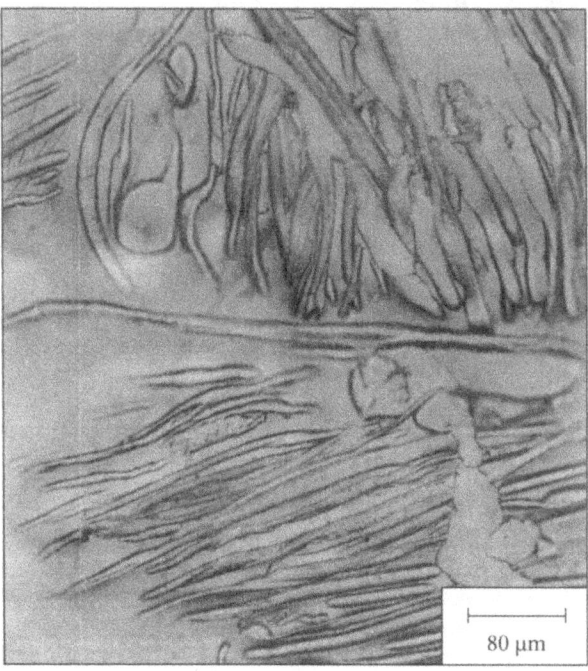

Fig. 138. Film imprint of a spotted area of the fabric from Fig. 137; a film-like coating from melted and flat-rolled polyester fibers can be clearly recognized.

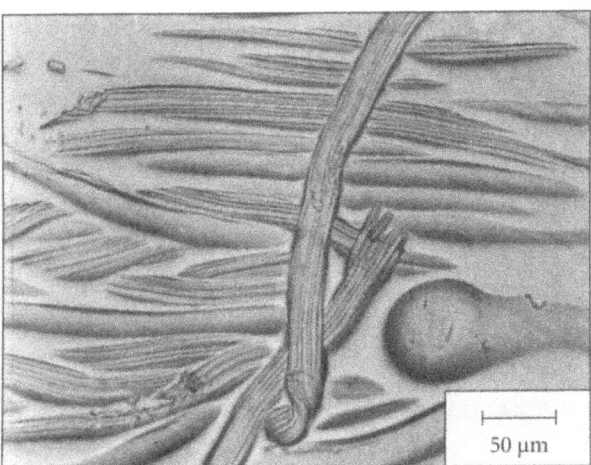

Fig. 139. Film imprint of a singed fabric made of polyester/viscose staple; typical melt ball at the end of a polyester fiber.

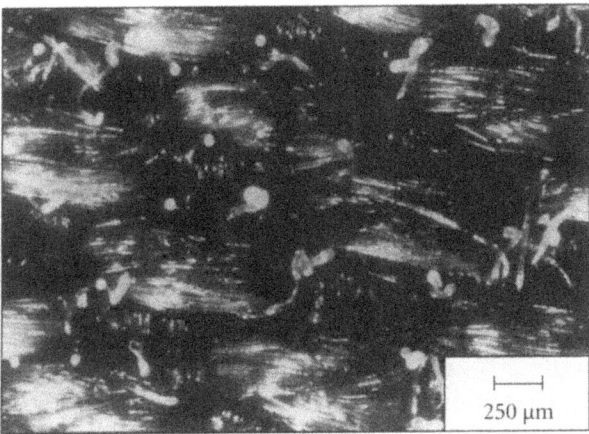

Fig. 140. Fabric made of polyester/viscose staple after a continuous dyeing process with undyed white polyester melt balls.

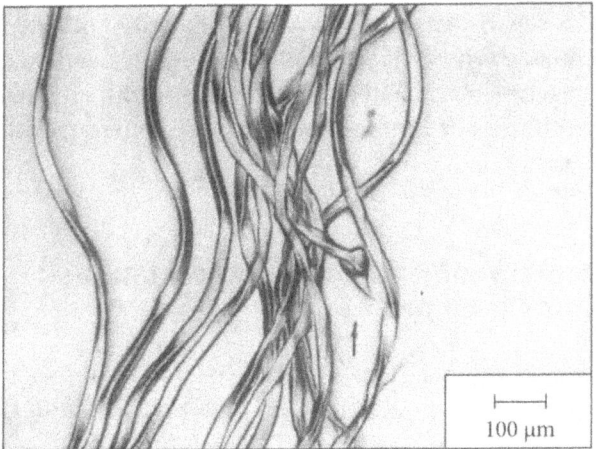

Fig. 141. Film imprint of a textured polyester weft yarn with a polyester fiber melted during singeing (arrow).

Fig. 140
Page 97
The hard melt balls were not dyed during the continuous dyeing process, Fig. 140. Therefore, light areas or streaks resulted in those areas in the fabric where the number of melt balls was particularly large. This at first seems to contradict the result of the previous example (chapter 4.1.6.2) where the polyester fibers are dyed darker. But it accords with general experience that the melt balls of the polyester fibers are dyed lighter than the intact fiber in continuous dyeing processes and darker in discontinuous dyeing processes. Therefore it is recommended and common in practise to singe fabrics made of synthetic fibers after dyeing.

4.1.6.4 Strength Loss After Singeing Due to Melting of Polyester Fibers – Practical Example

A fabric consisting of cotton in the warp and of a textured polyester filament yarn in the weft had partially lost its strength after finishing. The warp was dyed white and the weft black. The fabric had been singed, desized, washed and easy-care-finished.

In order to determine the cause, the cotton warp was examined by means of the pinhead reaction (chapter 2.3.2, Fig. 72–74), but no chemical damage to the cotton could be found. No mechanical damage was detectable on the cotton fibers. The cotton yarn had obviously not been damaged during finishing. Thus the damage had evidently been caused by the textured polyester yarn.

Fig. 141 – 142
Page 97/100
This yarn was isolated from the untreated material and the finished fabric and subsequently examined microscopically. Imprints on polypropylene film were taken because of the better depth of focus. Fig. 141 shows the weft yarn from the rejected fabric. In addition to the thermal deformation which results from texturizing according to the false twist method (corkscrew-like twists), a melted fiber can be clearly recognized. On the fabric replica, Fig. 142, the damage can be seen even more clearly. Numerous melted polyester fibers with melt balls can be recognized. The strength loss of the piece was therefore caused by inappropriate singeing of the fabric.

4.1.6.5 Streakiness After Singeing and Dyeing Due to Melted, Darker Dyed Polyamide Fibers – Practical Example

A fabric consisted of polyamide in the warp and of viscose staple in the weft. Due to marked streakiness in the direction of the weft it had been rejected. It was maintained that the damage was caused during weft yarn lubricating. This

deficiency could not be eliminated by stripping the dye and subsequent redyeing – in spite of thorough intermediate cleaning. Microscopic examination
showed that in the area of the dark streaks there were more thermally deformed
polyamide fibers than in the remaining piece. The material defect was therefore
located in the polyamide warp. The streak developed in the direction of the
flame tube which corresponded to the weft direction. Subsequent dyeing tests
showed that the melted polyamide fibers were dyed darker than the other
fibers, Fig. 143. The black and white photo also shows that the melted areas are
dyed darker. The fabric had undoubtedly been singed before dyeing.

Fig. 143
Page 100

4.1.7 Damage Caused by Ironing

Like the finisher, the garment manufacturer must also take into consideration
the thermal characteristics of synthetic fibrous materials during production.
However, this is only possible if he is aware of the fiber composition of the individual articles. Synthetic materials are particularly sensitive to overheating
during ironing. With the aid of temperature-controlled irons, ironing defects
can nowadays be avoided to a large extent. However, one practical example of
an ironing defect is to be explained in the following.

4.1.7.1 Light Stains on a Fabric Made of Polyester/Wool Caused by Ironing – Practical Example

Trousers made of polyester/wool showed light stains which looked like deposits
on the fiber. Even paints were suspected. The stains could be removed neither
by dry-cleaning nor by thorough stain removal. Microscopic examination
showed that the polyester fibers were thermally deformed in these areas; they
were rolled to form a film which caused the light stains on the fabric, thus
giving the impression of deposits, Fig. 144. The thermal deformation of the
fibers could only have been caused by local overheating. Since the head of the
iron was recognizable in some areas of the trousers, Fig. 145, these marks could
only have been produced during ironing.

Fig. 144–145
Page 100/101

4.2 Thermal Deformation of Synthetic Fibers Due to Frictional Heat

Overheating, either due to direct heat or due to friction, does not only lead to
the deformation of synthetic fibers; the fibers can become so soft that they

Fig. 146
Page 101

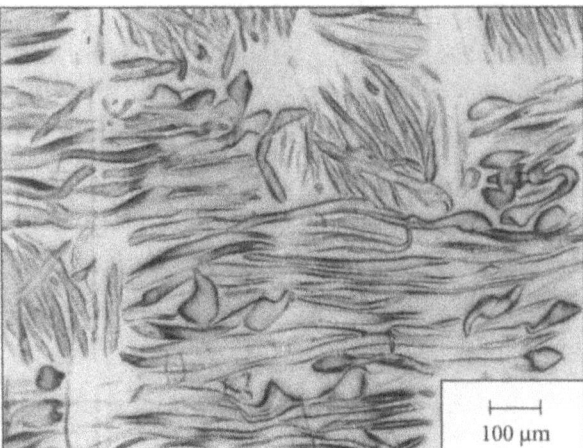

Fig. 142. Film imprint of the fabric as shown in Fig. 141 with numerous melted polyester fibers in the weft; warp made of cotton.

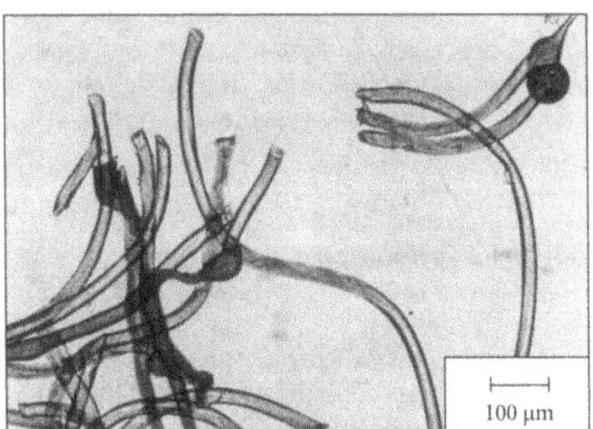

Fig. 143. Polyamide fibers, partially melted during singeing and thus dyed darker than intact fibers.

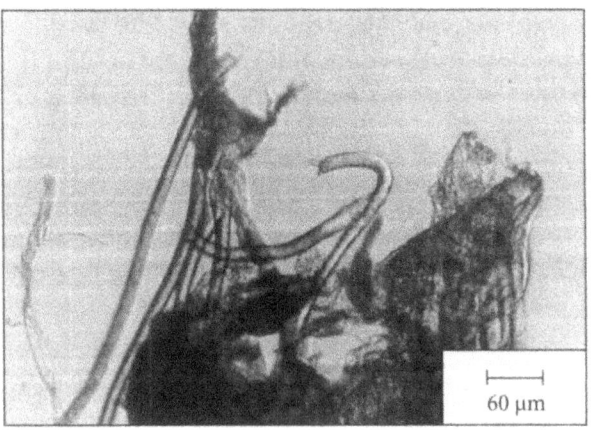

Fig. 144. Polyester fibers, rolled into a film during ironing due to overheating.

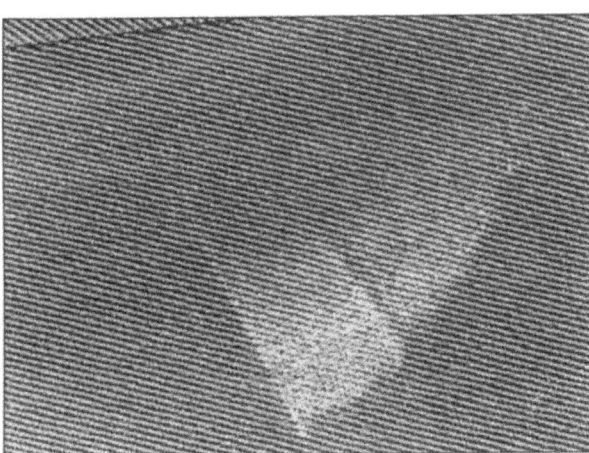

Fig. 145. Trousers made of polyester/wool with imprint of an iron.

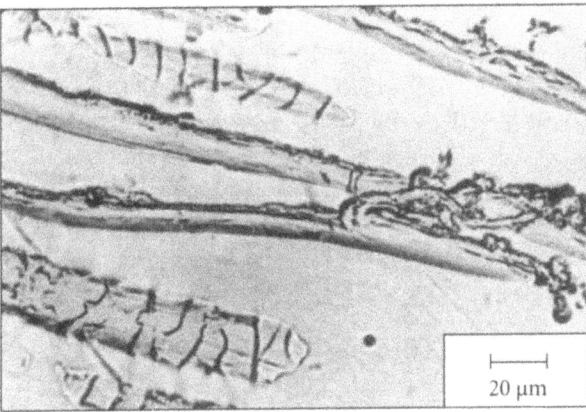

Fig. 146. Film imprint of a fabric made of polyester/wool with thermo-mechanical damage to the polyester fibers. Obviously, entire areas were torn from the outer skin of the fibers.

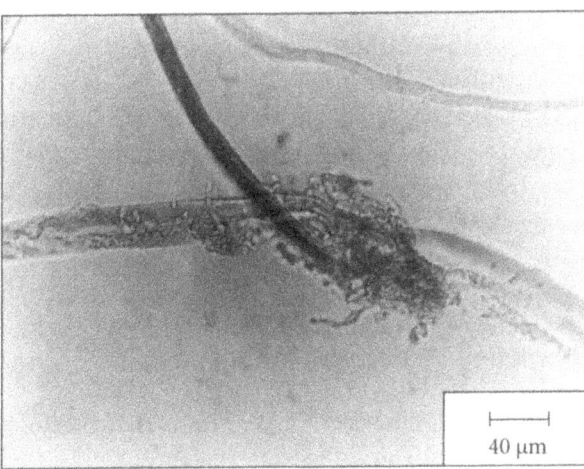

Fig. 147. Acrylic fiber with thermo-mechanical damage due to excessive friction against the thread guides and/or thread brakes during rewinding.

adhere to thread guides and rollers, and whole patches can be torn from the surface, Fig. 146. In the damaged areas, incident light is scattered diffusely so that the impression of light deposits or lighter-dyed stains results. The following practical examples illustrate this topic.

4.2.1 Streaks in a Piece of Knitwear Caused by Acrylic Fibers with Thermo-Mechanical Damage - Practical Example

Acrylic fibers have no melting point since they decompose before reaching the melting point. However, they first become soft and then sticky when heated.

Pieces of knitted fabric made of acrylic fibers were rejected because of their streakiness. The streaks in the fabric were exactly parallel to the threads. Since it was a yarn-dyed material, at first the streaks were thought to have been caused during dyeing.

Fig. 147
Page 101

However, microscopic examination showed that the fibers from the slightly lighter, mat streaks in the piece were in part abraded and flattened. As a result of the destruction of the roughened fiber surface they became optically lighter, Fig. 147.

A check in the plant revealed that the yarn had been insufficiently paraffin-treated, resulting in excessive friction.

4.2.2 Graying on a Dyed Acrylic Yarn - Practical Example

Fig. 148–149
Page 103

During dyeing of acrylic hanks, problems arose in the form of unlevelness within one hank as well as between the individual hanks. The unlevelness showed in the form of graying. In order to determine the cause of the damage, microscopic examination was performed. It showed that the surface of numerous acrylic fibers was abraded in the grayed fiber areas, Fig. 148. The piece of knitwear showed streaks parallel to the threads which were also visible on large film imprints because not only individual fibers were abraded but also the entire yarn, thus becoming more voluminous and hairy, Fig. 149. The defect can only have been caused by excessive friction during rewinding.

4.2.3 Light Patches on a Sewing Thread Made of Polyester Due to Thermo-Mechanical Damage - Practical Example

After dyeing and lubricating, light patches appeared on the surface of a sewing thread made of polyester; they looked like deposits. At first this defect was

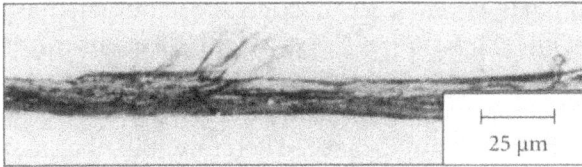

Fig. 148. Acrylic fiber with spliced and/or torn off fiber constituents from a grayed area of the yarn with thermo-mechanical damage.

Fig. 149. Film imprint of a piece of knitwear with streaks parallel to the threads which are caused by abrasion of the yarns during rewinding.

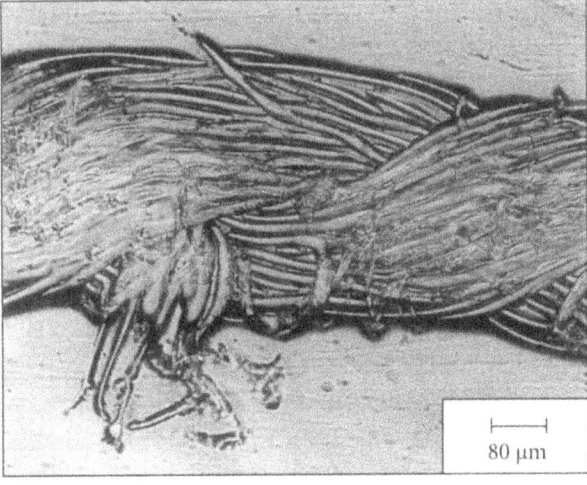

Fig. 150. Film imprint of a polyester thread. The polyester fibers are damaged thermo-mechanically by excessive friction on thread guides and thread brakes, thus giving the impression of light deposits.

Fig. 150
Page 103

attributed to the lubricant. However, since it could be eliminated neither through scouring nor through extraction of the yarn with different solvents, a film imprint was produced and examined microscopically. In the respective areas, thermally bonded fibers as well as abraded fiber particles were found, Fig. 150, causing an optically lighter appearance. Therefore, the fiber material had been damaged thermo-mechanically.

4.2.4 Thermal Deformation of Polyester Fibers as a Result of Excessive Spinning Speeds – Practical Example

Fig. 151–152
Page 105

During preparation of a blended yarn made of polyester/viscose staple, a large amount of dust developed on the ring spinning machines. Microscopic examination of the fiber dust showed that it mainly consisted of polyester, Fig. 151. The fibers had partially fused with each other and had been squeezed to form a fiber skin, Fig. 152. In addition, thermally deformed polyester fibers were also found in the yarn itself. Due to friction during the spinning process the polyester fibers had obviously been heated to such an extent that they were deformed, flattened, twisted and partially bonded to each other. The process was triggered by excessive spindle speeds.

4.2.5 Fiber Dust Formation During Twisting of a Polyester/Cotton Yarn – Practical Example

Fig. 153
Page 105

On two-for-one twisters the fibers are exposed to high friction. This leads to dust formation and heating of the synthetic fibers which can be so strong that they are deformed, broken and rolled into a thin fiber skin. Fig. 153 shows the micrograph of polyester fiber dust produced in this way which resulted from twisting of a yarn made of polyester/cotton. Such disturbances can be avoided by using the optimal production speed and suitable lubricants, which fulfill the high requirements with regard to surface smoothness, film formation, corrosion resistance, removability through scouring and bonding of the abraded obligomers.

4.2.6 Bar Formation Due to Polyester Fibers with Thermo-Mechanical Damage in a Fabric Made of Polyester/Wool – Practical Example

In a fabric made of polyester/wool, bar formation parallel to the weft was detected after scouring and dyeing. It was assumed that differences in the finish add-on were the cause. Extraction with petroleum ether yielded very

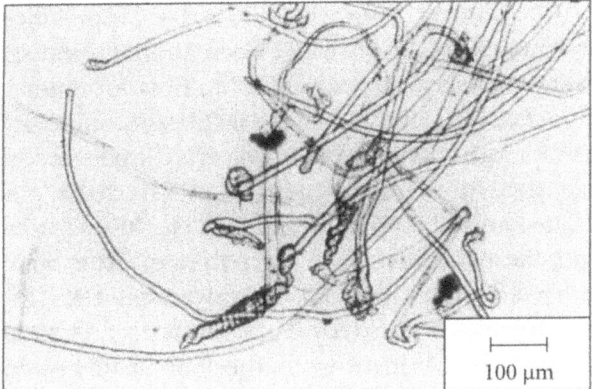

Fig. 151. Polyester fiber dust, resulting from excessive spindle speeds of a ring spinning machine.

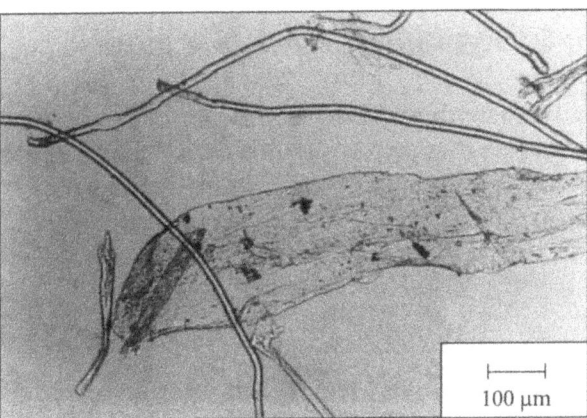

Fig. 152. Polyester fiber dust as in Fig. 151, partially squeezed to form a thin skin.

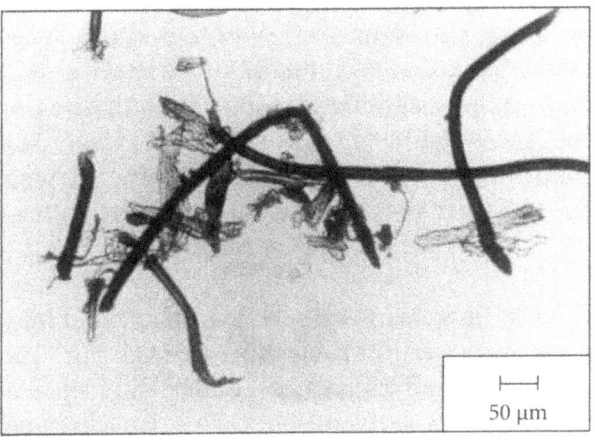

Fig. 153. Fiber dust, produced during twisting of a yarn made of polyester/cotton. Polyester fibers with thermo-mechanical damage which were partially rolled into a thin fiber skin can be recognized.

Fig. 154–155
Page 107
low values of between 0.01 and 0.02% for the extractable proportion. Since after extraction the weft bars remained with the same strength and shape as before, film imprints were taken of the barré and faultless areas for microscopic examination. On the imprints, the wool fibers could be easily distinguished from the smooth, structureless polyester fibers because of their characteristic scale structure. In the faultless areas of the piece the fiber material was clean and showed no sign of damage, Fig. 154. However, the warp threads were damaged at the weft bars. Here the polyester fibers were partially abraded or deformed while the wool fibers were faultless, Fig. 155. The bar formation was obviously caused by optical effects due to thermal deformation of the polyester fibers. Friction in the loom had heated them to such an extent that they were deformed.

4.3 Thermal Damage to Synthetic Fibers Due to Impact

The thermal deformability of synthetic fibers is also noticeable in the case of impact. Released heat energy leads to very pronounced and irreversible changes to the fibers. Here the so-called "shuttle marks" and "warp splashes" are of special importance; they occur frequently during processing of synthetic fibers on shuttle looms [41]. The following practical examples will illustrate the resulting problems.

4.3.1 Lighter Yarn Areas Due to "Shuttle Marks" on a Fabric Made of Acrylic Fibers – Practical Example

Fig. 156
Page 107
An upholstery fabric of acrylic yarn dyed dark blue was rejected because of light, obviously undyed yarn areas running in the weft direction over varying lengths of thread, usually for several centimeters. Since the yarn had been dyed on cylindrical cross-wound bobbins, it was at first assumed that the defect had been caused during dyeing. Preliminary examination with a magnifying glass showed that the color defect was restricted to the weft yarns. Fig. 156 shows a micrograph of the film imprint of a lighter yarn area under the microscope. It can be clearly seen that the fibers in the light, lustrous areas are flattened and stuck together.

Fig. 157–158
Page 109
Microscopic examination of fiber samples revealed that the fibers were split, and thin fiber skins were torn and/or squeezed off, Fig. 157. The squeezed-off fiber skins, Fig. 158, can be recognized quite easily. The flattened, thin, split, transparent fiber constituents are optically lighter and lustrous,

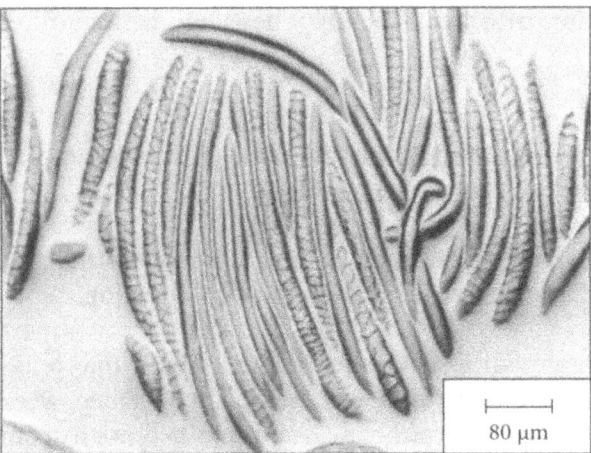

Fig. 154. Film imprint of a fabric made of polyester/wool without any sign of fiber damage.

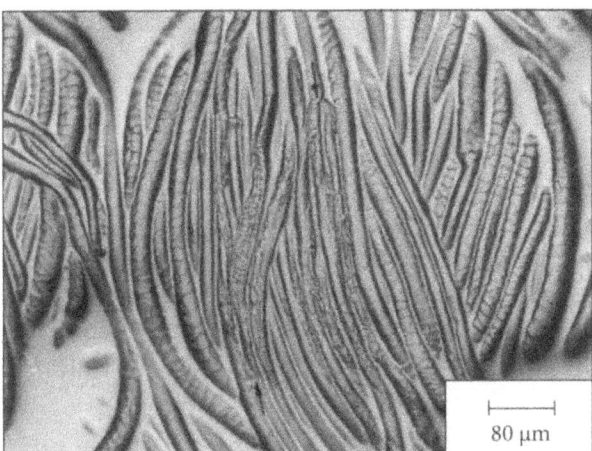

Fig. 155. Film imprint from the area of a weft bar of the fabric from Fig. 154. The wool fibers are intact but the polyester fibers in the warp show thermo-mechanical damage. The outer skin of the polyester fibers is in an abraded state.

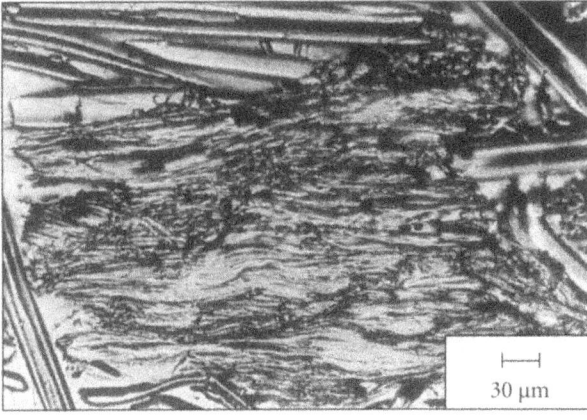

Fig. 156. Film imprint of an acrylic fabric with light, lustrous yarn areas in the weft. Acrylic fibers which are thermally deformed and flattened due to impact can be recognized.

thus giving the impression of a lighter color. These findings are typical of "shuttle marks".

4.3.2 Light Streaks in a Black Dyed Polyester Fabric Due to "Warp Splashes" – Practical Example

Fig. 159
Page 109
A black dyed polyester fabric showed small light streaks close to the right selvedge. As shown in Fig. 159, these streaks ran in the weft direction.

Fig. 160
Page 110
A preliminary examination with a stereo microscope showed that only warp threads caused this defect, the weft yarns were not involved. Fibers were then taken from the light areas of the warp threads and examined microscopically. In addition to flattened and bonded fibers, splitting and squeezed, thin, transparent fiber skins could be recognized, Fig. 160. Consequently, damage typical of impact could be observed [42]. Later it transpired that the fabric damage had been caused by shuttle stroke.

4.3.3 White Streaks in a Polyester/Acrylic Fabric Caused by "Warp Splashes" – Practical Example

Fig. 161
Page 110
Figure 161 shows a warp thread made of polyester/acrylic from a fabric which had been rejected because of small white streaks in the warp. From the figure it can be easily seen that the fibers are partially thermally bonded and flattened. The shuttle stroke partially squeezed them into thin, transparent fiber skins.

4.4 Thermal Deformation of Synthetic Fibers Due to Cutting, Punching and Sewing

During cutting and punching synthetic fibers can also heat up and thus become plastic. This leads to bonding and deformation at the cut ends and/or cut edges. This causes optical effects which can give the impression of color streaks. The degree of damage to the fibers depends on the finishing, the production rate and the quality of the blades used.

Another problem in the garment industry occurs during sewing of textile fabrics on high-speed sewing machines. As a result of mechanical friction between the sewing needle and the synthetic fibers, the needle can heat up in such a way that heat damage, fusion of the fibers and even loop damage can result. In order to prevent this damage, specific sewing thread lubricants are used.

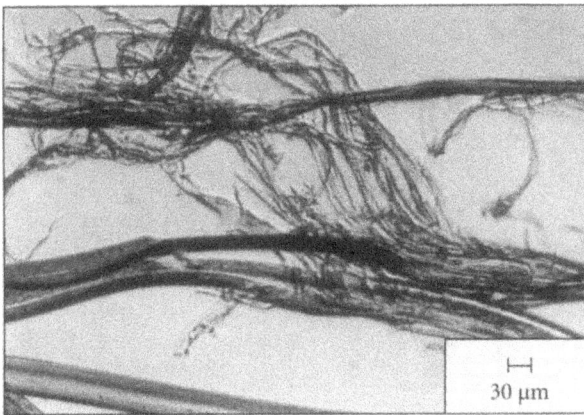

Fig. 157. Fiber preparation from the weft yarn of the fabric from Fig.156, with flattened and splintered fibers typical of "shuttle marks".

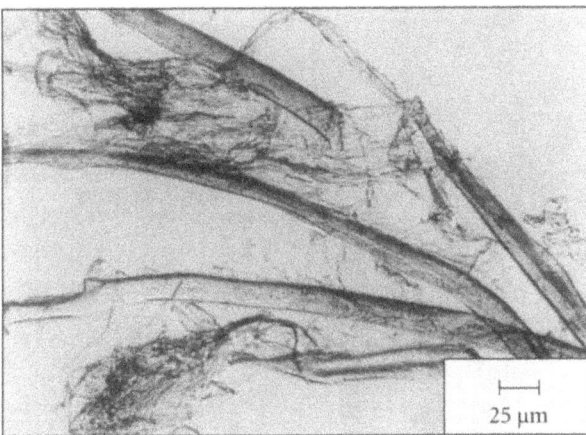

Fig. 158. Another fiber preparation of the yarn as shown in Fig. 156. These fibers, squeezed into thin skins, are also typical of "shuttle marks".

Fig. 159. Black dyed polyester fabric with light streaks in the weft direction. In this case so-called "warp splashes" caused by the shuttle stroke are responsible.

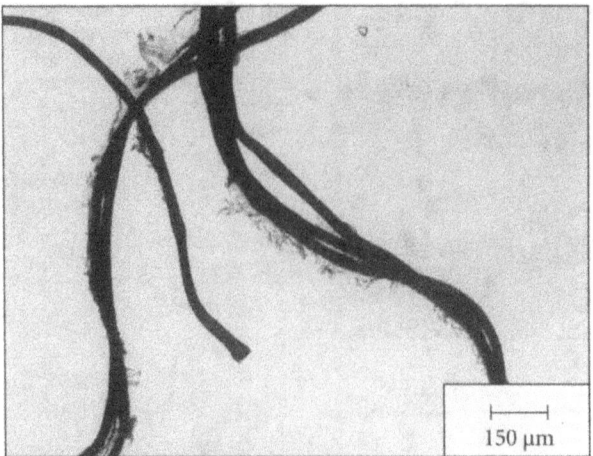

Fig. 160. Polyester fibers from the warp yarn of the fabric in Fig.159, flattened and thermally bonded by the shuttle stroke. Thin fiber skins are partially detectable.

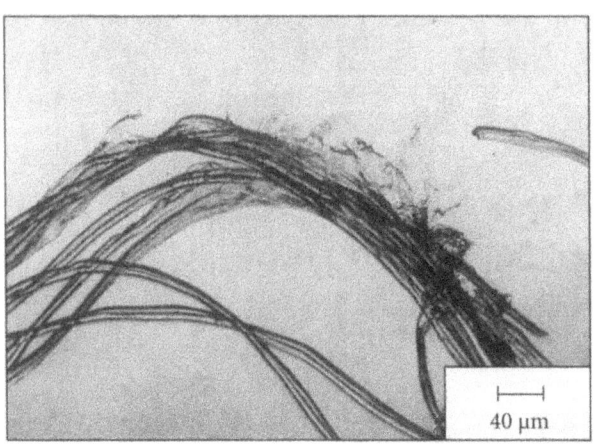

Fig. 161. The same damage as in Fig. 160, except that it is a fiber preparation made of an acrylic/polyester blended fabric.

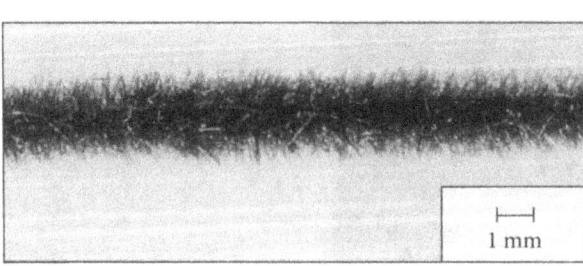

Fig. 162. Flocked yarn; core made of viscose, pile made of polyamide short staple.

4.4.1 Thermally Bonded Cut Ends in Polyamide Short Staple

Natural and man-made fibers (particularly polyamide fibers) are used for textile surface finishing by means of flocking, Fig. 162. Due to the high cutting sequence, cutting of the fiber cables for the production of polyamide short staple is difficult because of the thermal sensitivity of the material.

Fig. 162
Page 110

Microscopic examination of the raw flock has repeatedly shown that numerous fibers are partially thermally bonded, Fig. 163. The cut surfaces are not smooth, Fig. 164. The surfaces are compressed and partially resemble rivet heads. The compressed fiber ends are partially bonded. The separation of a thermally bonded flock is only possible through intensive mechanical processing during dyeing. With the aid of a specific softening agent, the separation process of the flock can be improved from the very beginning.

Fig. 163–164
Page 112

4.4.2 Streak Formation in a Plush Fabric Made of Acrylic Fibers
– Practical Example

Light and dark streaks parallel to the warp were found in a plush fabric whose pile consisted of acrylic fibers [43]. Microscopic examination showed that the constitution of the fiber ends of the pile differed greatly. In the light streaks the fiber ends were chopped, split and had partially formed thin fiber skins, Fig. 165; this – as has already been explained above – hints at thermo-mechanical damage. In the darker streaks, the fiber ends were mainly cut smoothly and were only occasionally slightly deformed, Fig. 166. From this it must be concluded that the fabric was sheared with defective blades, thus leading to streaks caused by optical effects.

Fig. 165–166
Page 112/113

4.4.3 Streak Formation in a Velour Carpet Made of Polyamide
– Practical Example

A velour carpet made of polyamide showed streaks. The streak formation could be recognized – although not very clearly – on large imprints of an extracted and a non-extracted sample. Therefore, the defect could neither be due to dyeing unlevelness nor to different preparation add-ons. Apparently optical effects were the cause of this material damage.

Several pile tufts were removed from the faultless areas and streaks and examined microscopically. In the faultless areas, the cut ends of the poly-

Fig. 163. Polyamide short staple, partially thermally bonded during cutting.

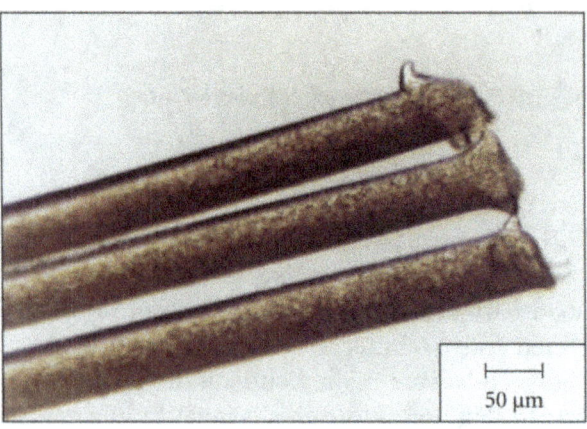

Fig. 164. Fiber preparation from the polyamide short staple in Fig. 163 with compressed fibers, partially bonded at the cut ends.

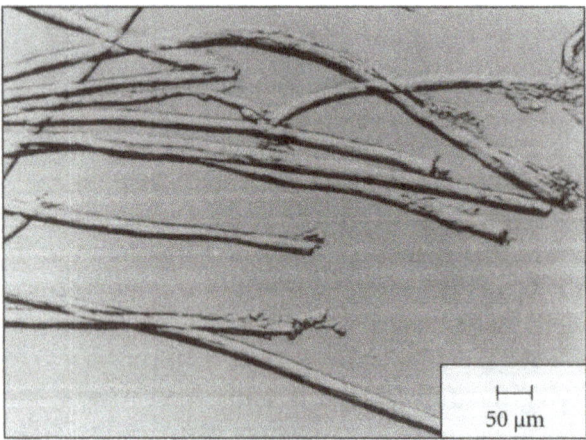

Fig. 165. Acrylic fibers from a light streak of a plush fabric. The cut ends are split and chopped and partially form thin fiber skins, thus causing optical effects in the form of light streaks.

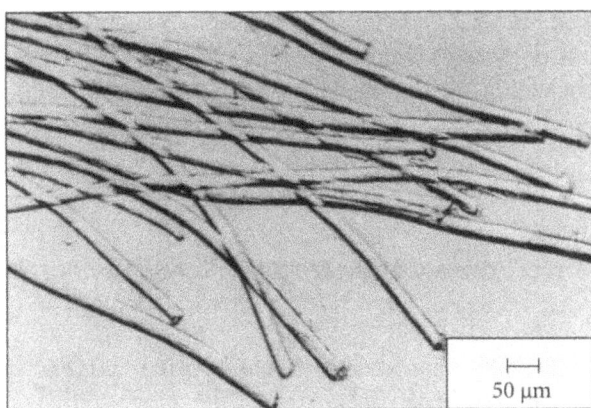

Fig. 166. Acrylic fibers from a dark streak of the plush fabric in Fig.165 with predominantly smooth cut ends.

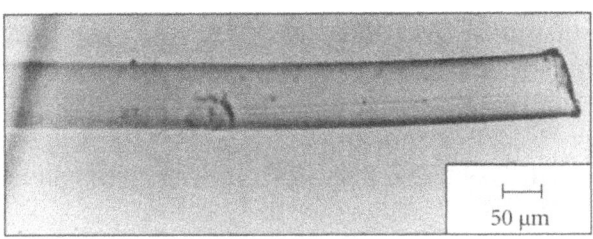

Fig. 167. Cut end of a polyamide fiber from a non-streaky area of a velour carpet.

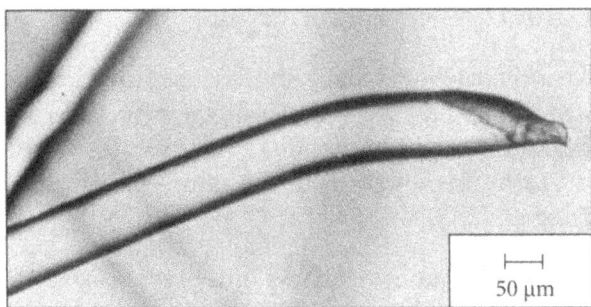

Fig. 168. Melted cut end of a polyamide fiber (drawn into a tip) from a grayed area of the velour carpet from Fig. 167.

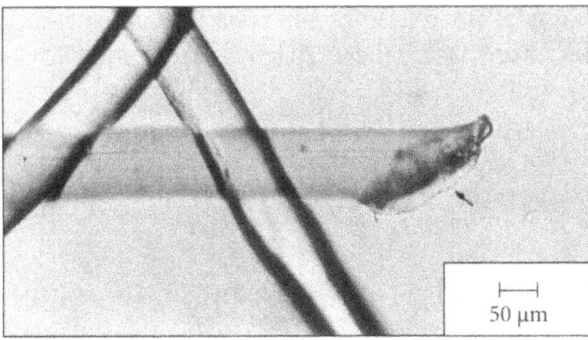

Fig. 169. Diagonally cut, slightly compressed end of a polyamide fiber from a grayed area of the velour carpet in Fig. 167, forming a thin skin.

Fig. 167–169
Page 113
amide fibers were smooth, Fig. 167. The cut ends of the fibers from the streaks were cut diagonally and under heat action drawn into a tip, Fig. 168. There were also diagonally cut fibers which were slightly compressed and formed a thin skin, Fig. 169. At these defective cut ends, incident light was scattered diffusely, thus causing graying.

4.4.4 Streak Formation in a Velour Carpet Made of Polypropylene – Practical Example

Fig. 170
Page 115
A velour carpet made of polypropylene was rejected because of streakiness. It showed more or less strongly developed light areas.

Microscopic examination showed that the fibers had partially melted at the cut ends and formed a thin, transparent fiber skin, Fig. 170. The cause was inappropriate cutting of the loops.

4.4.5 Bonding of Punched Pieces of a Knitted Fabric Made of Polyamide – Practical Example

Pieces of knitted fabric made of polyamide stuck together after punching. This was attributed to electrostatic charging of the material.

The test on conductivity of the fabrics actually showed that they had no anti-static finish. But even after discharging of the material, the parts stuck together, especially at the cut edges. Thus the bonding of the pieces was not caused by electrostatic charge; apart from this, pieces with the same charge would repel each other and not stick together.

Fig. 171
Page 115
Microscopic examination of the film imprints of different cut edges showed that the polyamide fibers were compressed or squeezed in these areas. Some fibers had the appearance of rivet heads. The compressed fiber ends were occasionally slightly bonded. Apparently the fibers had been heated up during cutting with blunt blades, thus causing thermal deformation and bonds at their cut ends, Fig. 171.

4.4.6 Detection of Cutting Defects on Polyester Fibers Through Dyeing of the Cut Ends

The dyeing of polyester fiber ends with m-cresol/Fat Red as described in chapter 2.4.4 is also suitable for the detection of cutting defects during pro-

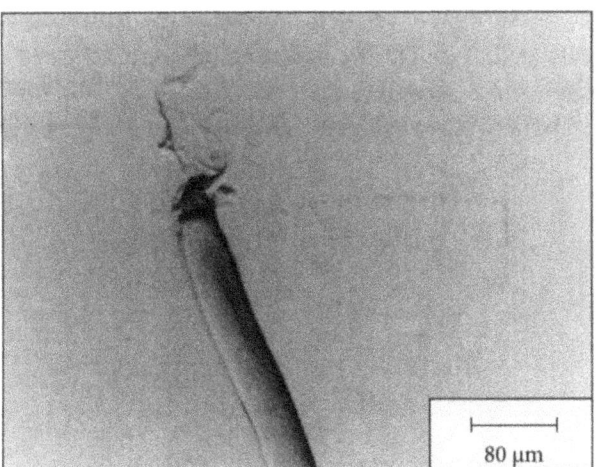

Fig. 170. Cut end of a polypropylene fiber with thermomechanical deformation from a light streak of a velour carpet.

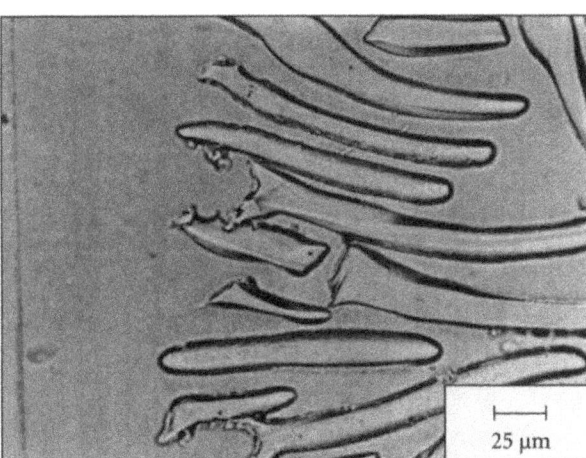

Fig. 171. Strongly magnified film imprint of a punched polyamide fabric. At the cut ends, squeezed, compressed and slightly bonded polyamide fibers can be recognized easily.

Fig. 172. Cross-section of a velour fabric from polyester/cotton, dyed with m-cresol/Fat Red. The squeezed cut ends of the polyester fibers are dyed intensively red.

Fig. 172
Page 115
duction. Smoothly cut fiber ends are only marginally dyed, while squeezed fiber ends are dyed intensively red. Figure 172 shows a cross-section of a velour fabric made of polyester/cotton after dyeing of the fiber ends. The ends of the polyester fibers which were deformed during cutting are dyed intensively red (see Fig. 92).

5 Streaks and Bars in Textile Fabrics Due to Yarn Differences and Technological Reasons

Differences in the yarn count, yarn twist and plytwist create different yarn volumes. They are generally referred to as yarn differences and are noticeable as streaks or bars parallel to the threads in woven fabrics and knit-wear. They must be attributed to purely optical effects. During examination in reflected light they are caused by the fact that adjacent, more voluminous yarns reflect light to a greater extent. The observer thus gains the impression that there are lighter streaks in these areas than in those with larger gaps between the threads. Here, light is reflected to a lesser extent, thus giving the impression of a darker dyeing or a darker color streak. In transmitted light, conditions are exactly the opposite. This is schematically illustrated in Fig. 173 and 174. However, luster differences between more or less intensively twisted yarns can lead to differences in lightness [44]. There are several possibilities for macroscopic and/or microscopic detection of yarn differences:

Fig. 173–174
Page 118

- Examination of the streaky textile fabric in transmitted light under the stereo microscope [5, 43, 45],
- Microscopic examination of the isolated yarns [46],
- Preparation of imprints of the textile fabrics or the isolated yarns [6, 7, 8, 10, 47].

It is more difficult to ascertain the cause of streaks and bars in pile goods. This is not least due to the fact that large film imprints are difficult to produce. From such imprints, unlevelness in dyeing could be quite easily distinguished from structural differences and deposits. Only with a very short pile is this feasible in pile goods. Otherwise, so-called cross-sections have to be prepared for microscopic examination of the material. For this purpose the pile fabrics have to be cut parallel to the threads; they are then locked into a cardboard frame or glued onto black or white cardboard so that the cross-section of the fabric can be examined with a magnifying glass and/or low-magnification microscope.

The examination of buckled surfaces from a side view is very similar to the examination of cross-sections.

Fig. 173. Schematical representation of the effects of yarn differences during examination of textile fabrics *in reflected light.* The thinner yarn optically gives the impression of a darker color streak.

Fig. 174. The same situation as illustrated in Fig. 173, *in transmitted light.* Under these conditions the thinner yarn optically gives the impression of a light color streak.

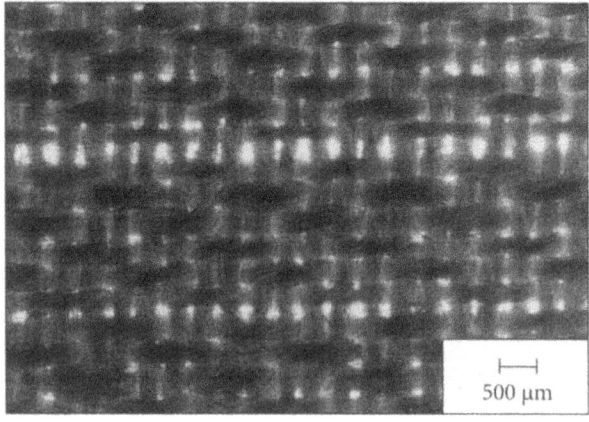

500 μm

Fig. 175. Streaky viscose fabric in transmitted light. Yarn differences in the spinning twist and plytwist as well as different distances between the warp threads cause optical effects which are visible as streaks.

In another test method, the individual pile tufts are isolated along the light and dark streaks and/or bars and are checked for irregularities under the microscope.

5.1 Streaks Due to Variations in the Yarn Volume or Yarn Count

5.1.1 Streaks Parallel to the Threads in a Viscose Staple Fabric – Practical Example

After dyeing, streaks parallel to the warp could be recognized in a viscose fabric. First, inappropriate desizing of the warp was considered to be the cause; however, after stripping and cleaning of the fabric the textile defect still had the same extent and form as before. The magnified micrograph of the fabric in transmitted light, Fig. 175, clearly showed that considerable differences occurred in the warp with respect to the spinning twist and plytwist of the yarns. This must have been the cause of the streaks.

Fig. 175
Page 118

Similar causes of the defect were also detected on an overcoat poplin fabric, Fig. 176, a closely woven cotton fabric with yarn irregularities that appeared only within relatively short thread lengths (Fig. 177 and 178), and a decorative fabric with viscose staple warp and viscose filament weft, Fig 179 and 180.

Fig. 176–180
Page 120/121

5.1.2 Warp-Streaky Twill Due to Differences in the Yarn Count – Practical Example

A twill made of pure cotton yarn was rejected because of its marked warp streakiness. It was a yarn-dyed material with only the warp being dyed but not the weft. Therefore, the dyehouse was blamed for the defect. No differences in color could be seen on the removed warp threads. Microscopic examination revealed considerable yarn differences in the warp. Figure 181 shows the fabric in transmitted light under the stereo microscope. With a relatively weak magnification the yarn differences can already be easily recognized. Streakiness is especially apparent due to the contrast between the dyed warp and white weft.

Fig. 181
Page 121

5.1.3 Streak Formation in a Tubular Knitted Fabric, Made of Mercerized Cotton Yarn, Due to Differences in the Twisting Effects – Practical Example

Tubular knitted fabric made of pure cotton yarn clearly displayed repeated streaks after dyeing. The distance between the streaks was approx. 1.3 cm. The

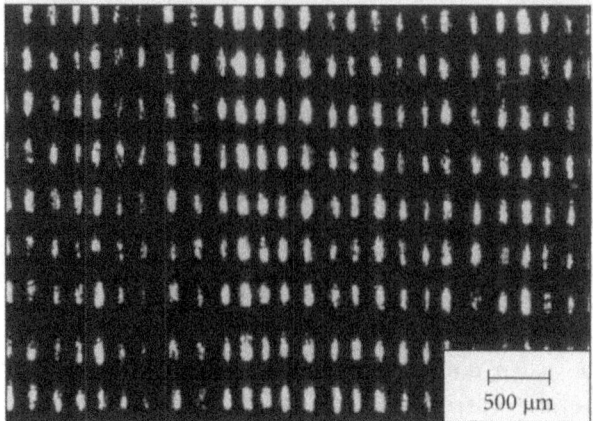

Fig. 176. Overcoat poplin in transmitted light. Yarn differences in the warp cause optical effects noticeable as streaks.

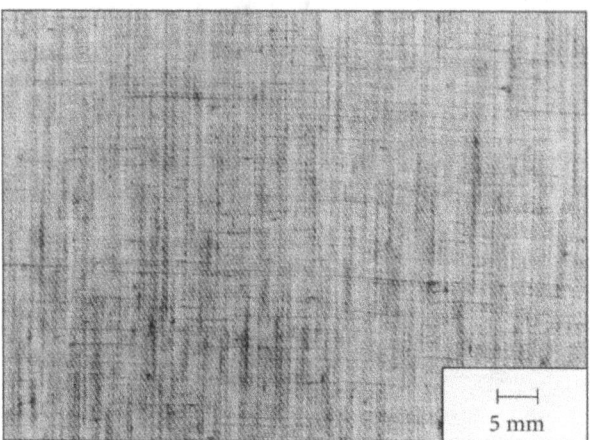

Fig. 177. Cotton graycloth in transmitted light. Large yarn differences within short thread lengths cause streakiness after dyeing.

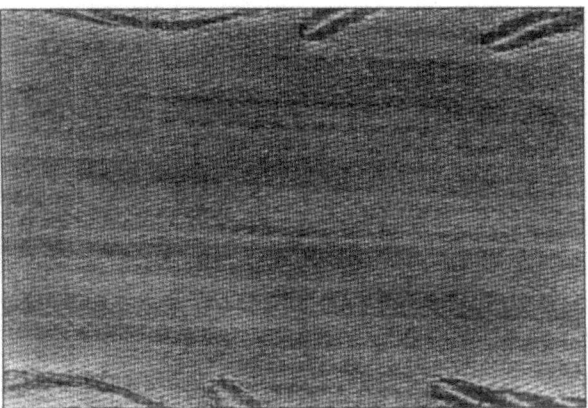

Fig. 178. The same fabric as in Fig. 177 in transmitted light, however with insignificant yarn differences; therefore, there is no streakiness.

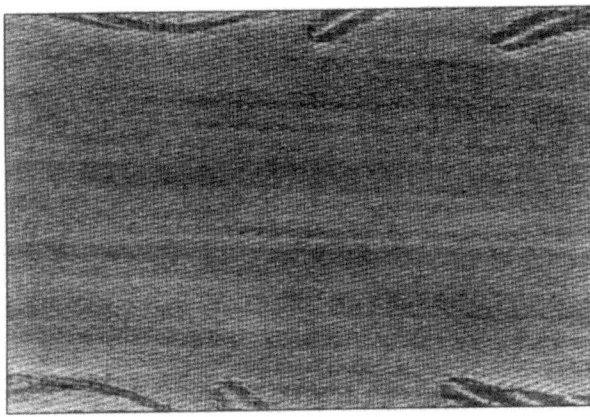

Fig. 179. Transmitted light micrograph of a decorative fabric with streaks in the warp.

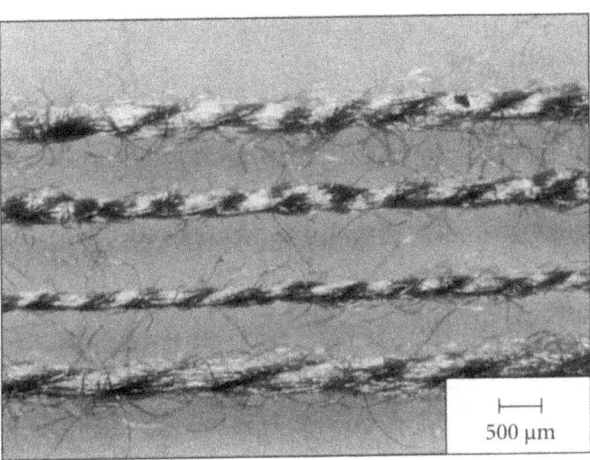

500 μm

Fig. 180. Warp threads from the streaky decorative fabric in Fig. 179. Differences in the spinning twist and plytwist account for the streak formation.

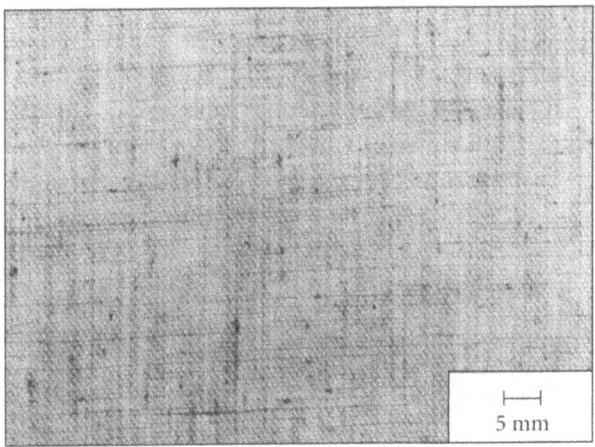

5 mm

Fig. 181. Streaky cotton twill in transmitted light. Even with a relatively weak magnification, the yarn differences can be easily recognized in the warp (running from the left to the right).

cotton yarn was singed, mercerized, knitted in an unbleached state and then dyed. Streak formation in the tubular knitted fabric was first explained by differences in mercerisation. However, microscopic examination of cotton fibers from streaks and from the remaining piece led to the conclusion that there were neither differences in mercerization of the yarns (cf. chapter 5.3.1, Fig. 208 and 209) nor fiber damage.

Fig. 182–183
Page 123 On a large imprint of the fabric, streaks could be clearly recognized, Fig. 182. The streaks could not even be removed by stripping the dye, thorough intermediate cleaning and redyeing. The streak formation had to be attributed to differences in the structure of the yarns, which was proven by microscopic examination. There were differences in the plytwist, Fig. 183.

5.1.4 Streaks and Bars in Cotton Fabrics Due to Varying Hairiness of the Weft Yarn – Practical Example

Fabrics made of pure cotton were streaky in the weft direction. In the rejected fabric pieces, the warp was dyed blue, the weft was white. Preliminary inspection of the fabric elucidated that the streaks and/or bars were also visible in transmitted light. This hinted at structural differences, namely at a different volume of the yarns. The bars ran from one edge to the other in a slight curve but still remained parallel to the threads due to a distortion of the weft yarns.

In order to determine the cause of this material fault, a large film imprint was prepared from a barré fabric section. The interfacial line (light/dark) could be easily recognized on both the imprint of a sample that was grease-free after extraction and a non-extracted sample.

Fig. 184
Page 123 Weft yarns were isolated from light and dark bars and film imprints were prepared. Microscopic examination yielded differences in the hairiness of the weft yarn, Fig. 184. Different yarn twists could not be found. The weft yarn taken from the light bars was obviously more roughened than the remaining yarn. The blue-dyed warp threads were covered by the white fibers of the weft yarn protruding from the yarn bundle, thus causing an optically lighter shade.

5.1.5. Streak Formation Due to Differently Twisted Mouliné Yarns – Practical Example

A fashionable men's suiting made of pure wool was rejected because of streakiness in the warp direction. A CMC size which had not been totally

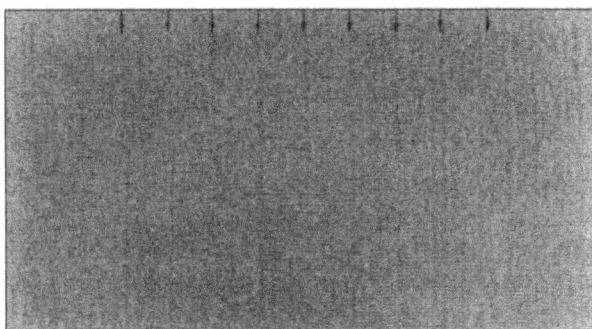

Fig. 182. Film imprint of tubular knitted fabric made of cotton with streak formation (arrows).

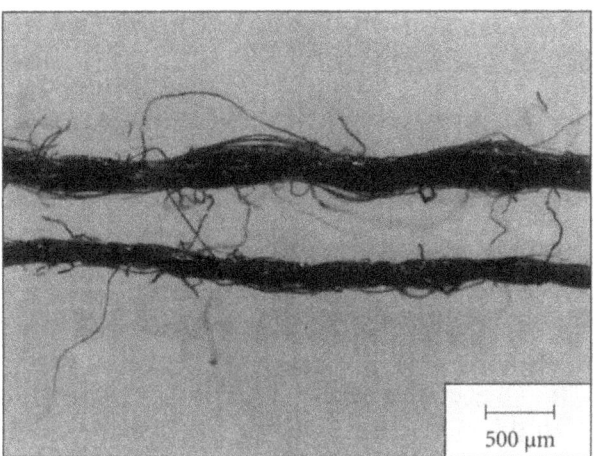

Fig. 183. Cotton yarns from the tubular knitted fabric with different twist turn in Fig. 182.

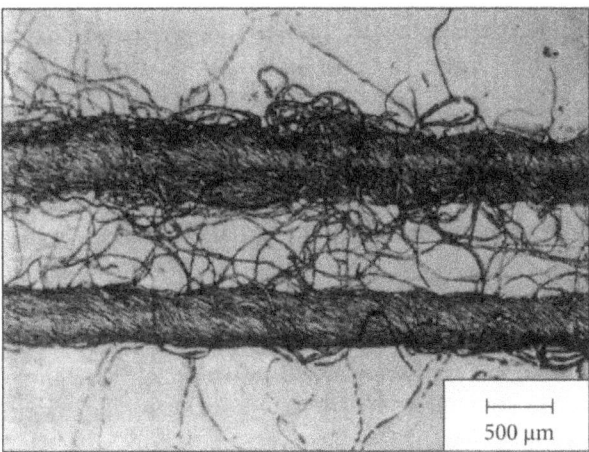

Fig. 184. Film imprints of weft yarns from a cotton fabric with streaks and bars. Above: Roughened and hairy yarn from a lighter bar. Below: Yarn from a darker bar with considerably reduced hairiness.

washed out was considered to be the cause of this. However, even after thorough scouring the defect still had the same intensity.

Fig. 185
Page 125
A fabric cross-section in the weft direction of the rejected fabric was prepared. It revealed that the fibers were not soiled, Fig. 185; incrustations caused by the sizing agent are not detectable on the warp threads. It can also be seen that different-colored mouliné yarns were processed in the warp.

Fig. 186
Page 125
Microscopic examination of the mouliné yarns revealed that they displayed considerable differences in twisting – even if only within short thread lengths, Fig. 186.

5.1.6 Warp Streakiness in a Polyester/Wool Fabric Due to Differences in Yarn Twist – Practical Example

A fabric section made of 50/50 polyester/wool showed marked warp streakiness. First, it was ensured that the warp was free of size.

Fig. 187 – 188
Page 125/126
Large film imprints revealed warp streakiness both in the original fabric and in the extracted, grease-free sample. Consequently, dyeing unlevelness due to oil or grease impurities and/or as a result of fiber finish residues could not be the cause; therefore, structural differences within the warp threads had to account for the defect. Film imprints of the warp threads clearly showed differences in yarn twist, Fig. 187 and 188. They can be easily recognized by the angle of the fibers to the yarn axis. With this method, it is easy to determine whether the volume differences in the yarn must be attributed to differences in the yarn count or in the yarn twist.

5.1.7 Streakiness in a Piece of Knitted Fabric Made of Bulked Acrylic Yarn Due to Differences in Volume and/or in Bulking – Practical Example

Pieces of knitted fabric (yarn-dyed acrylic fibers) were rejected because of the streaky and skittery appearance of the goods.

Fig. 189
Page 126
Streakiness was visible not just in normal daylight but also during examination of the knitted fabric in transmitted light, Fig. 189. This circumstance suggests differences in the volume of the yarns.

Fig. 190
Page 126
Figure 190 shows the yarns which had been isolated from the piece of knitted fabric. The differences in the volume and/or in the bulk of the yarns – which account for the streak formation – can be clearly recognized.

Fig. 185. Fabric cross-section of a pure-wool men's suiting made of mouliné yarns with warp streakiness.

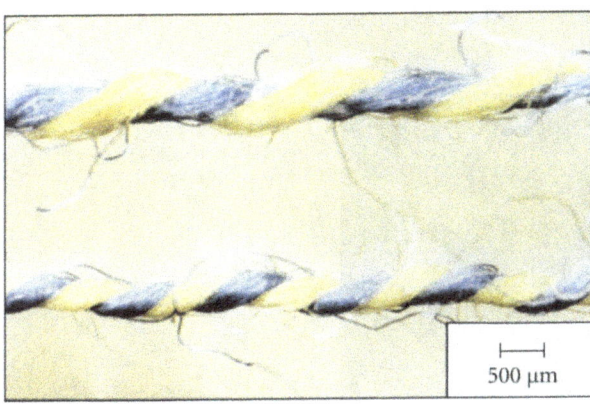

Fig. 186. Mouliné yarns with different spinning twist and plytwist from the fabric in Fig. 185.

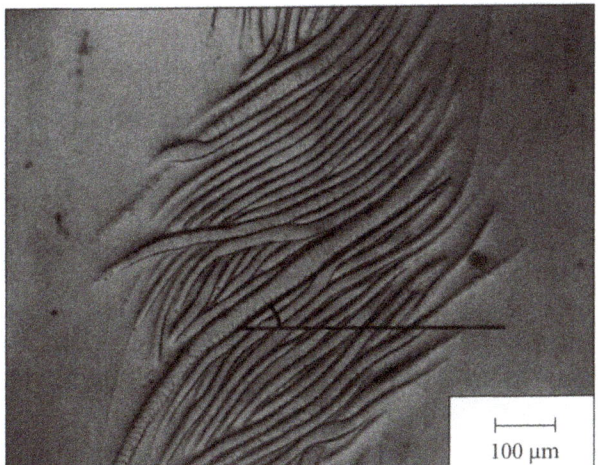

Fig. 187. Film imprint of a yarn from a polyester/wool fabric with warp streaks. The yarn is more strongly twisted than the yarn in Fig. 188, which can be deduced from the angle of the fibers to the yarn axis.

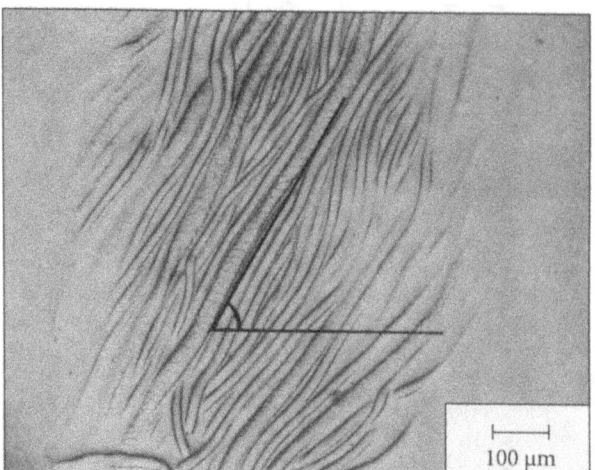

Fig. 188. Film imprint of a yarn taken from another area of the fabric corresponding to that in Fig. 187. The yarn is less strongly twisted.

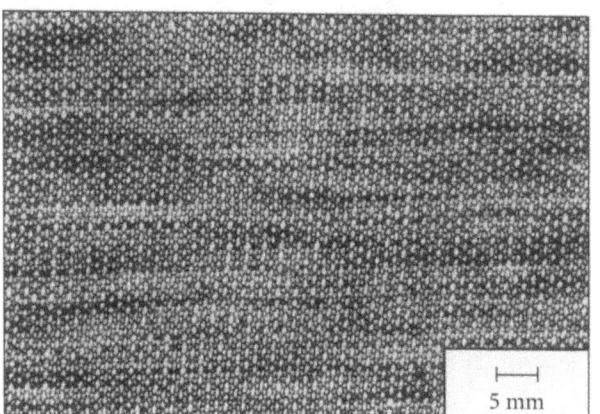

Fig. 189. Heavily streaked piece of knitted fabric made of bulked acrylic yarn in transmitted light.

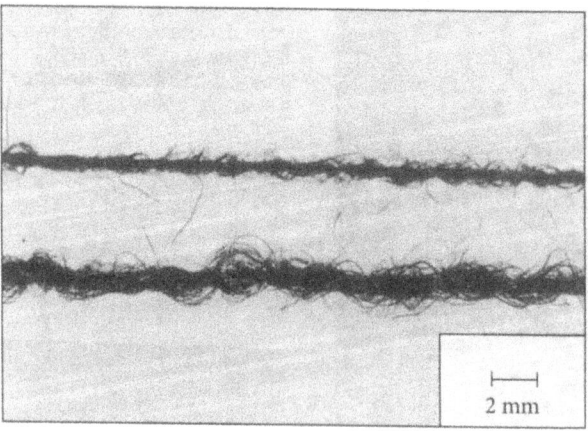

Fig. 190. Bulked acrylic yarn with different volume and/or bulk from the streaky piece of knitted fabric in Fig. 189.

In addition to streakiness, raised and depressed areas could be recognized on the piece of knitted fabric, Fig. 191. The yarns must have been stretched differently in the piece of knitted fabric. This must be attributed to a varying thread unwinding force during knitting. Later, it became apparent that the yarn had been unevenly paraffin-treated.

Fig. 191
Page 128

5.2 Streaks and Bars Parallel to Threads Due to Yarn Mixture Errors

Often, yarn errors cannot be recognized in the untreated material. This is normally the case when differences only occur in the fiber blend while yarn count as well as spinning twist and plytwist are equal. Only after dyeing do they become apparent. This also applies to cases where yarns were processed from fibers of different origin. The following examples will show how these mistakes can be recognized.

5.2.1 Dark Weft Bar in a Cotton Fabric After Dyeing – Practical Example

There was a brown-dyed fabric made of cotton which contained a darker weft bar, Fig. 192. Differences in the hand of the fabric were also noticeable between the normally dyed part and that with the darker weft bar. The darker weft bar felt harder than the remaining piece. The weft yarn allegedly was from a single spinning lot.

Fig. 192
Page 128

By means of microscopic examination pure cotton yarn was found in the weft of the normally dyed fabric part, Fig. 193, while there was a cotton/viscose staple blended yarn in the darker weft bar, Fig. 194. Consequently, weft yarns had been mixed up in the weaving mill.

Fig. 193–194
Page 128/129

5.2.2 Weft Bars in a Fabric Made of Wool/Viscose Staple – Practical Example

Fabric made of wool/viscose staple which showed weft bars with different color shades after dyeing.

Dyeing tests on pieces of gray-state fabric with Neocarmin W helped to clarify the bar formation. Figure 195 shows wool fibers and viscose staple fibers from the weft yarn of a bar. With Neocarmin W wool was dyed uniformly yellow;

Fig. 195
Page 129

Fig. 191. Streaky piece of knitted fabric with raised and depressed areas resulting from an uneven thread unwinding force during knitting due to an uneven application of paraffin to the yarn.

Fig. 192. Fabric made of allegedly pure cotton. Dark weft bar (left) after dyeing due to a yarn mixture error.

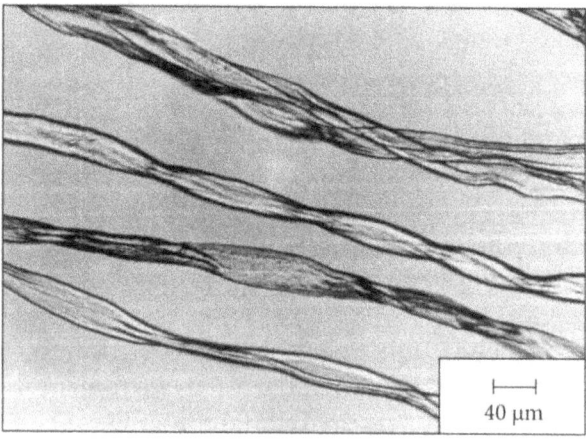

Fig. 193. Fiber sample of the weft yarn from the normally dyed part of the fabric in Fig. 192. The weft yarn consists of pure cotton.

40 µm

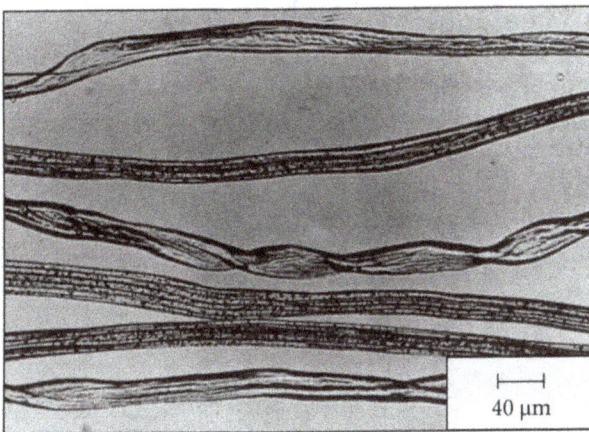

Fig. 194. Fiber preparation of the weft yarn from the darker bar in the piece in Fig. 192. It is a blended yarn made of cotton/viscose staple.

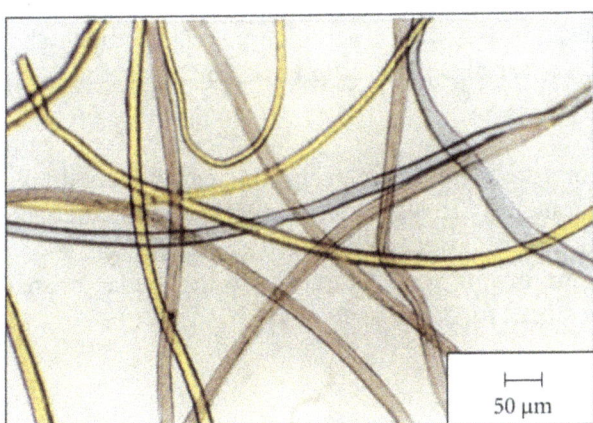

Fig. 195. Fiber preparation from a weft barred blended fabric. Wool, dyed yellow with Neocarmin W, and viscose staple, partially dyed pink, partially dyed blue by this reagent. There are two different types of viscose staple.

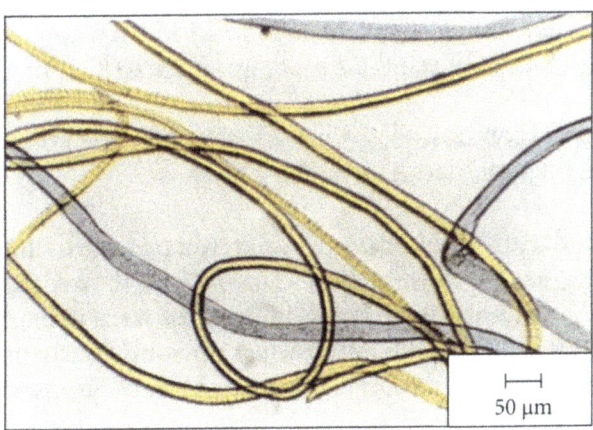

Fig. 196. Fiber preparation from another part of the fabric in Fig. 195. Wool dyed yellow with Neocarmin W, and viscose staple, only dyed blue by the reagent. There is only one type of viscose staple here.

Fig. 196
Page 129
viscose staple fibers were dyed blue or pink. Therefore, two different viscose staple types were processed together in a yarn mixture. The pink dyed viscose staple fiber had a dogbone cross-section – recognizable from the streaks in the longitudinal view. The blue-dyed viscose staple fiber had a round cross-section. In another bar, however, only one viscose staple type (dyed blue by Neocarmin W) could be detected in the weft yarn, Fig. 196.

Again, weft yarns had quite obviously been mixed up in the weaving mill.

5.2.3 Streaks in an Acrylic Fiber Fabric Due to Yarns of Different Origin – Practical Example

Fig. 197–198
Page 131
Acrylic fibers are produced by dry or wet spinning processes. Depending on the processing method, their fiber cross-sections have different shapes. The fiber cross-section can be circular, Fig. 197, dogbone, crenelated, bean-shaped or dumb-bell, Fig. 198. There are also large differences in the dyeing behavior of the fibers.

Fig. 199
Page 131
Figure 199 shows a decorative fabric made of acrylic fibers which, after dyeing, showed a streakiness parallel to the warp. The defect could not be recognized in the undyed material. Examination of the fiber cross-sections showed that acrylic yarns of different origin with different dyeing behavior had been processed, thus causing streak formation.

5.2.4 Warp Streaks in Polyamide Fabrics Due to Yarn Mixture Errors – Practical Example

A fabric sample made of polyamide showed light streaks in a fixed repeat parallel to the warp. The distance between the streaks was approx. 85 mm.

Examination of the material composition revealed that polyamide 6.6 had been processed both in the lighter and darker dyed warp threads.

Fig. 200–201
Page 132
Therefore, lighter and darker dyed warp threads were isolated and film imprints were prepared. Microscopic examination clearly showed that the titer of the individual filaments was finer in the lighter dyed warp yarns than in the darker dyed yarns, Fig. 200 and 201. The defect must obviously be attributed to a yarn mixture error which could not be recognized in the undyed material.

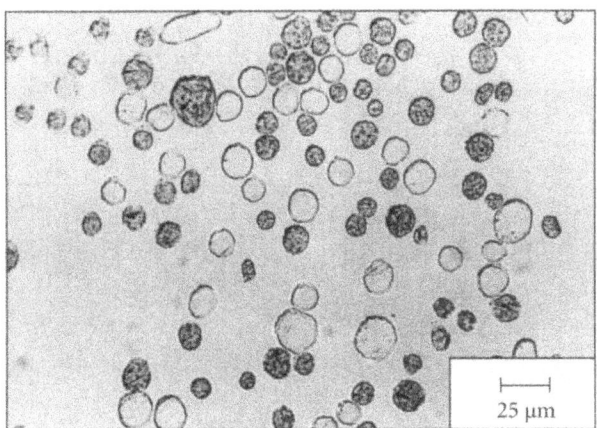

Fig. 197. Lustrous and dull acrylic fibers with round cross-section.

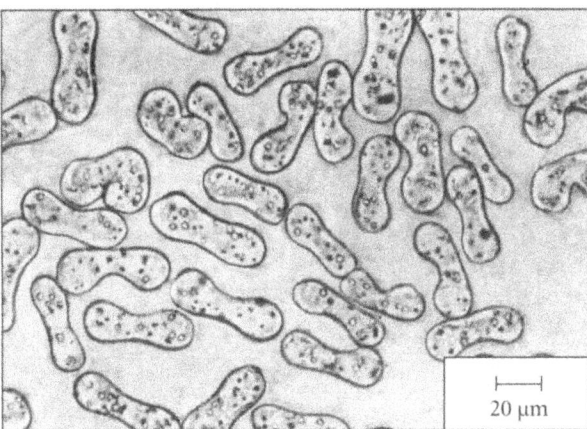

Fig. 198. Dull acrylic fibers with dumb-bell cross-section.

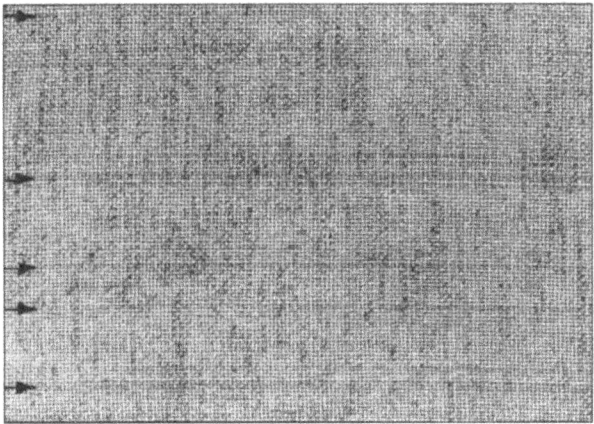

Fig. 199. Decorative fabric made of acrylic with warp streaks after dyeing due to processing of acrylic fibers of different origin.

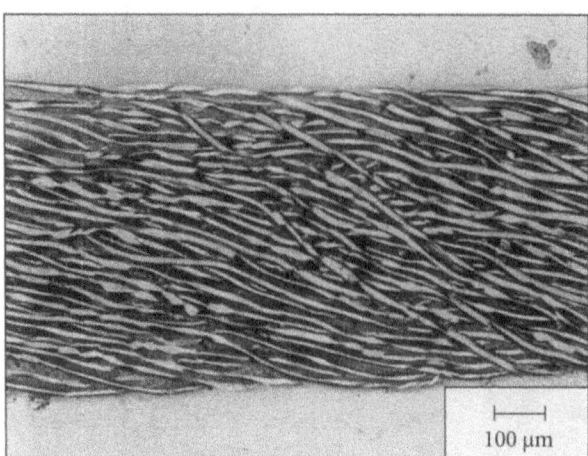

Fig. 200. Film imprint of a textured lighter dyed warp thread from a polyamide fabric with warp streakiness.

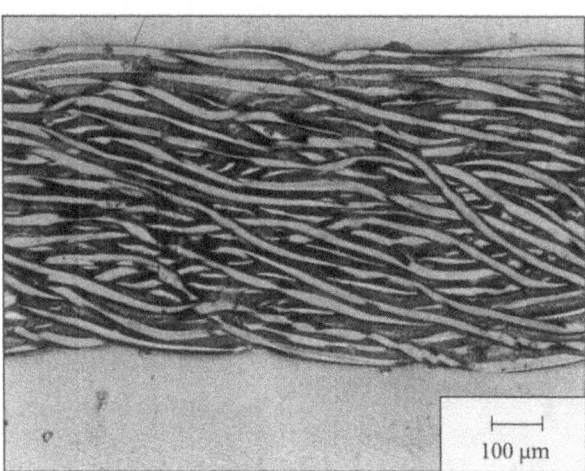

Fig. 201. Film imprint of a warp thread from a darker dyed fabric comparable to Fig. 200. The filaments have a coarser titer.

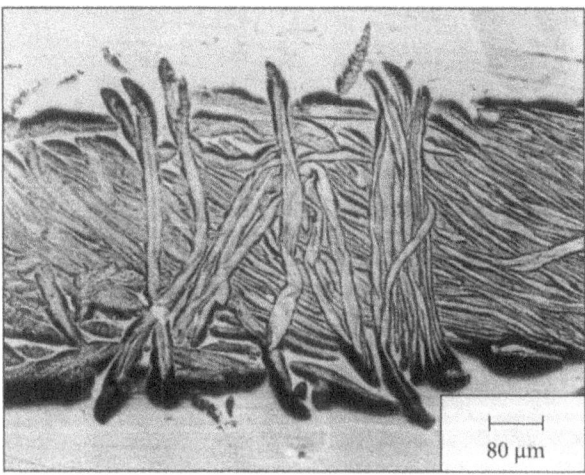

Fig. 202. Film imprint of an OE rotor yarn made of cotton with typical wrapped fibers.

5.2.5 Dye Unlevelness in Cotton Pieces Due to Yarns Manufactured According to Different Spinning Processes – Practical Example

Trousers were manufactured from bleached cotton fabrics which were dyed with substantive dyes after garment manufacture. After the dyeing process, several pairs of trousers were dyed in the desired color shade, while others had a slightly more yellow cast.

Examination of the material composition showed that all pieces were made of pure cotton. The pinhead reaction did not reveal any chemical damage to the cotton fibers, cf. 2.3.2, Fig. 72.

Further examination centered on the yarns themselves. Weft and warp yarns were isolated from the brown and yellowish sections and imprinted on poly-propylene films for comparison. Microscopic examination revealed that yarns had been used which were manufactured using different spinning processes. The brown dyed pieces were made of a yarn which had been manufactured according to the OE rotor spinning process. The typical wrapped fibers serve as an example, Fig. 202. The yarns from the yellowish pieces were produced according to the ring spinning process, Fig. 203.

Fig. 202–203
Page
132/134

5.2.6 Warp Streaks and Bars Due to Yarns of Different Cotton Origin – Practical Example

A black-dyed cotton fabric showed streaks and bars parallel to the warp which were slightly lighter than the remaining fabric. Their development was at first attributed to size residue; however, this could not be detected.

A streaky section was then stripped, cleaned intermediately and dyed again. This did not influence the state of the fabric.

Subsequent examination of the material composition of the warp threads proved that pure cotton was processed as specified by the manufacturer. No signs of chemical damage could be recognized.

Film imprints were prepared of the warp yarns from the streaky and the non-streaky portion of the fabric. Microscopic examination showed that there are more immature and dead cotton fibers in the lightly dyed warp threads than in the properly dyed warp threads, Fig. 204 and 205, cf. chapter 2.3.1. The un-treated material, which was only later available, was also examined. The red/green test on the untreated material, Fig. 67, confirmed that yarns from

Fig 204–205
Page 134

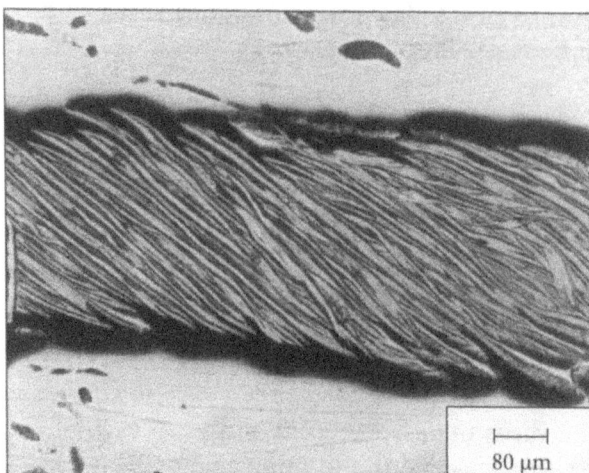

Fig. 203. Film imprint of a cotton ring-spun yarn.

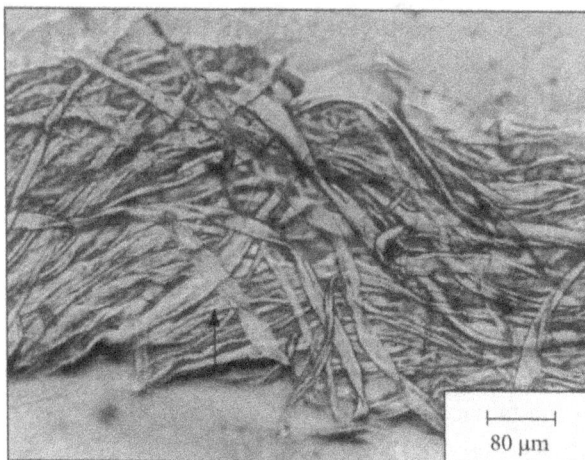

Fig. 204. Film imprint of a lightly dyed cotton warp yarn from a streaky fabric with a greater number of immature and dead cotton fibers. Due to their thin cell wall the dead cotton fibers are broad and transparent (arrow). An accumulation of these fibers leads to random orientation at the yarn surface.

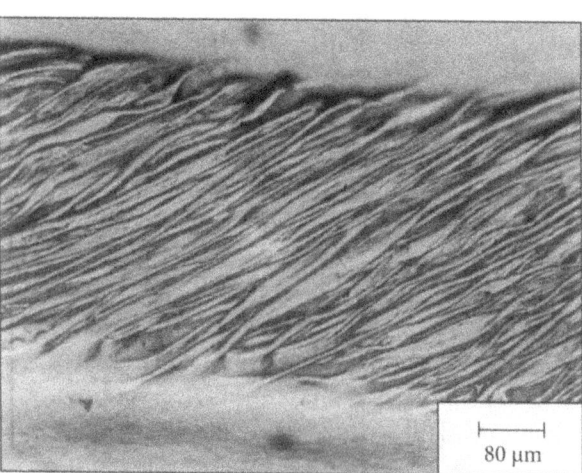

Fig. 205. Film imprint of a warp yarn dyed darker than the fabric in Fig. 204 with mature cotton fibers.

different spinning lots with cotton fibers of different maturity had been processed into one piece.

5.2.7 Warp Streaks in a Black-Dyed Wool Fabric – Practical Example

A black-dyed wool fabric displayed a streak parallel to the warp which could also be recognized in transmitted light.

Microscopic examination of the isolated warp thread confirmed that there were no differences in the fiber composition; this thread was also made of pure wool.

An imprint of the isolated yarn on a polypropylene film revealed clear differences in volume, hairiness and yarn twist as compared to a yarn sample from a non-streaky area, Fig. 206. Thus it was a clear yarn mixture error.

Fig. 206
Page 136

5.2.8 Color Differences and Streaks in Ribbons Made of Secondary Acetate from Different Spinning Lots – Practical Example

After finishing, pink-dyed acetate ribbons seemed lighter, duller and streakier than the standard fabric. This was attributed to finishing; it was assumed that the acetate filament had been saponified as a result of the effect of alkali.

Neither during the dyeing test with Sirius Red 4B, nor during microscopic examination of the fibers immersed into zinc chloride-iodine solution, could any signs of a surface saponification of the acetate fibers be found, see chapter 2.4.5.

Examination of fiber cross-sections which were subsequently prepared (Fig. 207) showed that the cross-section from the dull streaks was a finer lobe-shape than that of fibers taken from the shiny streaks. The fact that acetate filament from different spinning lots had been processed accounted for the defect.

Fig. 207
Page 136

5.3 Streak and Bar Formation Due to Other Yarn-Related Influences

Apart from yarn differences, other errors in the textile manufacturing process can lead to the formation of streaks and bars parallel to the threads. This is explained by means of practical examples.

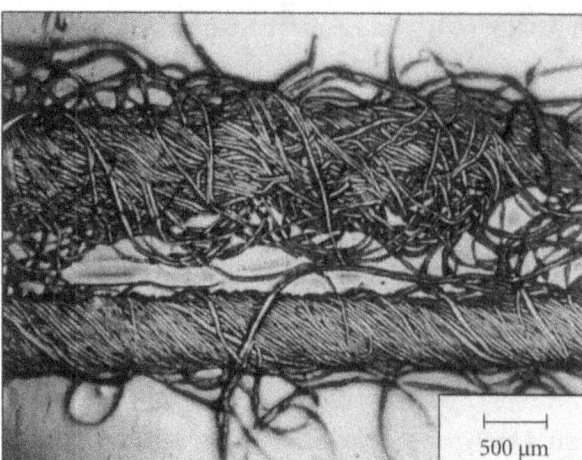

Fig. 206. Film imprint of wool yarns which show differences in volume, twist and hairiness. The yarn differences are the cause of streakiness.

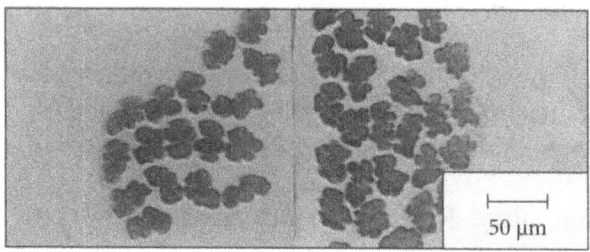

Fig. 207. Cross-section of secondary acetate fibers from different spinning lots. Left: Cross-section is coarsely lobed. Right: Cross-section is finely lobed.

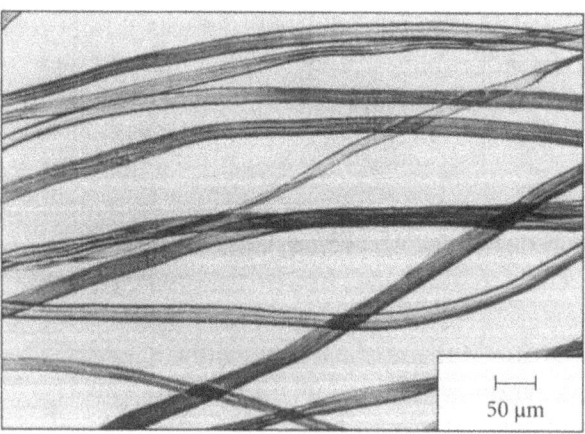

Fig. 208. Properly mercerized cotton. The fibers are smooth, roller-shaped and have no structure.

5.3.1 Streaks Due to Incorrect Mercerization – Practical Example

A dark-blue dyed cotton interlock fabric was heavily streaked; the streakiness could not be eliminated by stripping the dye and redyeing.

Figure 208 shows cotton fibers from darker dyed streaks in the piece. The fibers no longer display twistings and distortions typical of cotton; they are roller-shaped, which means that the fibers are properly mercerized. Figure 209 displays cotton fibers from lightly dyed streaks which are not properly mercerized. The typical twists and distortions of untreated cotton still occasionally occur, see chapter 2.3.1, Fig. 69 and 70.

Fig. 208 – 209
Page 136/138

The streak formation was caused by differences in the mercerizing effect.

5.3.2 Streaks Due to Differences in the Blend – Practical Example

Fabric samples of blended polyester/wool 55/45-yarns contained darker streaks and/or bars in the weft.

Threads were removed from both the faultless areas and the dark streaks and were microscopically examined; this illustrated that only some of the wool fibers were dyed black. The remaining wool and polyester fibers were white. The weft yarns in the dark bars contained a greater number of black dyed wool fibers than those of the faultless fabric, Fig. 210. Consequently, the cause of the streak and/or bar formation could be easily ascertained by microscopic examination. It was undoubtedly due to differences in the blend.

Fig. 210
Page 138

5.3.3 Weft Streaks in a Polyamide Fabric Due to Absence of Protective Twist – Practical Example

In a fabric with weft and warp made of polyamide 6.6, there were, after dyeing, small, lighter thread pieces within short thread lengths.

This textile damage - also referred to as "snow formation" - was also visible on a large film imprint, Fig. 211. Therefore, it could not be due to dyeing un-levelness.

Fig. 211
Page 138

Microscopic examination of the film imprint revealed that the yarn was untwisted; in the lighter thread pieces the weft was completely flat, Fig. 212.

Fig. 212 – 213
Page 139

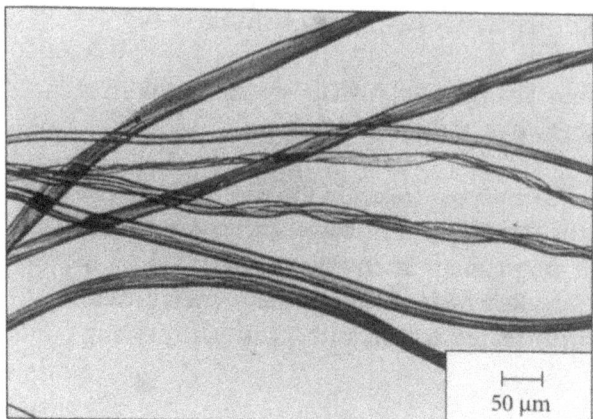

Fig. 209. Imperfectly mercerized cotton from a light streak of a dark-blue dyed cotton interlock fabric. The twists of non-mercerized cotton are still detectable with individual fibers.

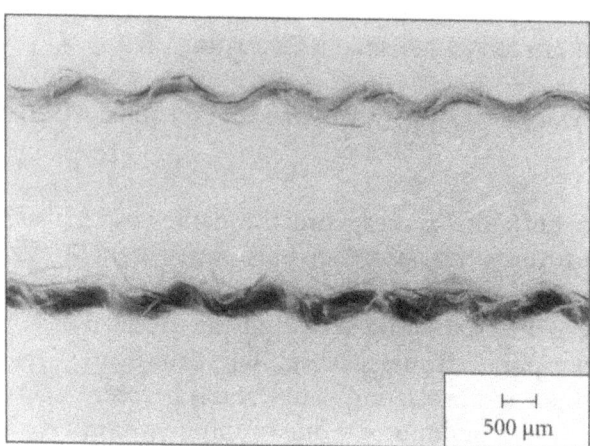

Fig. 210. Melange yarn made of polyester/wool. Deviations in the melange cause weft bars and streaks.

Fig. 211. Film imprint of a polyamide fabric with streaks from left to right in the weft within short thread lengths.

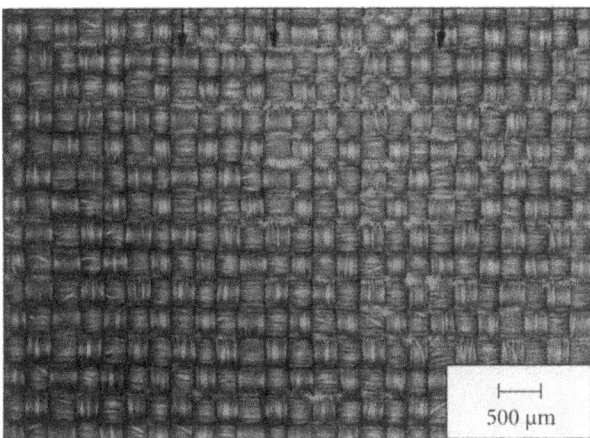

Fig. 212. Section from the film imprint in Fig. 211. In the untwisted weft yarn the filaments – running from top to bottom – are parallel within shorter thread lengths. The threads are lighter and have a better luster due to optical effects.

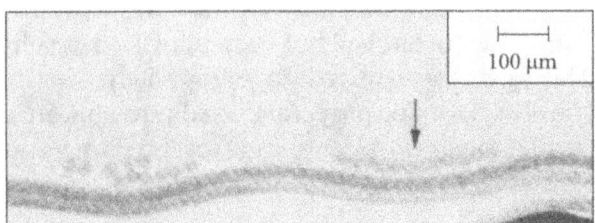

Fig. 213. Fabric cross-section of the polyamide fabric in Fig. 211. The filaments of the untwisted weft yarn are sometimes parallel (arrow), sometimes on top of each other.

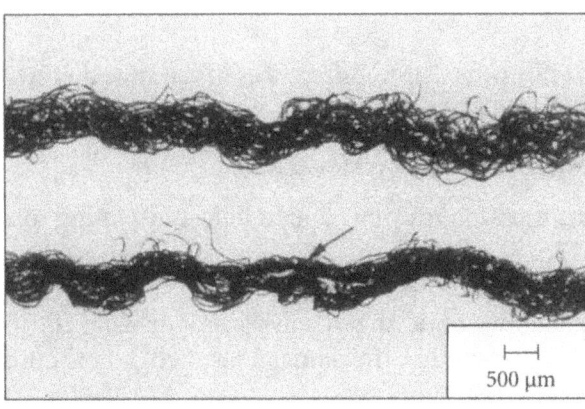

Fig. 214. Texturizing differences in a polyester yarn. Above: Normal bulked yarn. Below: One thread in a twoply yarn is in places not untwisted (arrow), therefore there is no bulk.

This is clearly illustrated by the fabric cross-section, Fig. 213. The individual fibers are parallel. In these flat yarn areas, incident light was strongly reflected, and as a result optical lightening and/or glossy spots were produced.

This damage could already be seen on the imprint of the untreated material. This could have been avoided through the use of protective twist in the weft yarns.

5.3.4 Warp Streaks Due to Incorrect Texturing – Practical Example

A fabric with a warp made of textured polyester yarn displayed streakiness parallel to the warp. The yarn was texturized by means of the false twist method, using a coning oil which was believed to account for the streak formation. The dry cleaned pieces displayed the same defect after dyeing.

Fig. 214
Page 139
Darker and lighter warp threads were carefully separated from the fabric and microscopically examined. It was found that the warp yarn was a twoply yarn. In the case of the darker yarn, one thread had not been untwisted, resulting in reduced bulk, Fig. 214. This yarn was texturized inappropriately. Dark streaks in the yarn bundle could be observed. The properly bulked yarn appeared to be more lightly dyed since incident light is more strongly scattered by the greater crimp of this yarn.

5.4 Streaks in Pile Goods

There are various causes of streaks and bars in pile goods. The following reasons are frequently found:

– different opening of the pile tufts due to differences in the spinning twist and plytwist of the yarns,

– differences in the length of cut and/or pile tuft length,

– ends of fiber of the tufts not smoothly cut due to defective cutting tools, see chapter 4.4, Fig. 165–172.

In the latter case, using microscopic fiber analysis it is usually difficult to make absolute statements as to whether the damage resulted during cutting or shearing.

5.4.1 Streaks Parallel to the Threads in a Tufted Carpet Made of Pure Wool, Caused by a Deeper Set Tuft Row – Practical Example

In a tufted carpet made of pure wool, a dark streak parallel to the pile could be seen more or less clearly, depending on the incidence of light; in UV light, however, this streak was not visible.

Pile tufts were removed from streaky areas and their residual grease content was determined by means of extraction; no difference could be detected as compared to the remaining pile yarn. There were also no differences in dye shade or blend.

Cross-sections were then prepared from different areas of the carpet and examined microscopically. This examination revealed that a tuft row was incorporated more deeply in the area of the streak. Here, the tufts were shorter than those in the remaining piece. Depending on the incidence of light, a shadow was cast on the deeper tuft row, thus causing a dark streak, Fig. 215.

Fig. 215
Page 142

5.4.2 Streaks Parallel to The Threads in a Tufted Carpet Made of Pure Wool Due to Different Needling – Practical Example

A tufted carpet made of pure wool was rejected because of streakiness parallel to the threads; this irregularity was caused by one or more tuft rows.

To explain this, a large number of tufts were removed from the light and dark streaks and examined. Neither dyeing variation between the tufts of the light and dark streaks nor differences in the residual grease content could be detected through petroleum ether extraction.

Microscopic examination of the cross-section of a streaky area showed that the distance between the individual tuft rows varied considerably, Fig. 216. In the densely needle-punched areas, the tufts could not open. This led to the formation of a darker streak since on these tufts incident light was not as well dispersed as on the tufts of intact areas.

Fig. 216
Page 142

The streaks in the carpet were therefore produced by an optical effect caused by differences in the spacing of the needle rows.

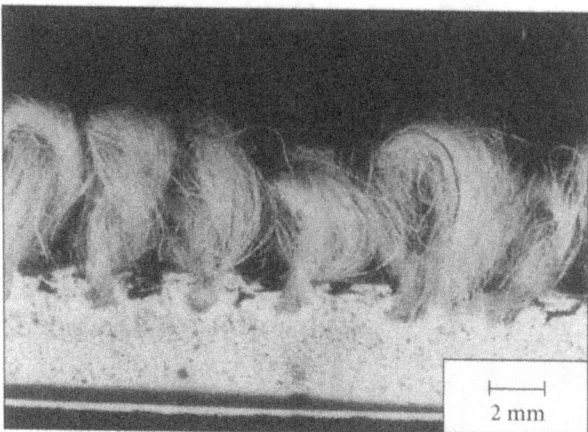

Fig. 215. Tufted carpet made of pure wool, with a more deeply incorporated tuft row, which causes streakiness.

Fig. 216. Streak formation due to different spacing of the tuft rows in a tufted carpet. Unopened tufts are found in the more densely needle-punched areas in the right micrograph section.

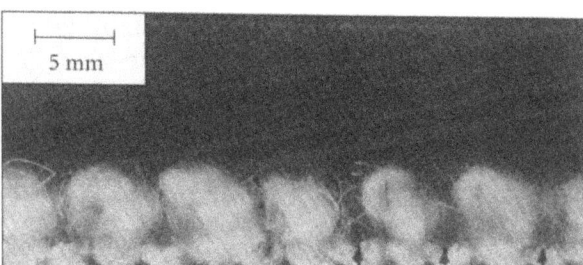

Fig. 217. Cross-section of a streaky woven wall-to-wall carpet. Left: Slightly twisted, open tufts; resulting in a continuous surface. Right: More strongly twisted, closed tufts; light penetrates through to the base (arrows), conveying an impression of color streaks.

5.4.3 Streak Formation in a Woven Wall-to-Wall Carpet Due to Yarn Differences – Practical Example

A woven carpet whose pile was made of 700 tex × 2 white wool yarn was rejected because of streakiness and barré parallel to the threads. This problem was attributed to deficient scouring of the wool yarn and it was assumed that the residual grease content varied within the yarn. However, with 0.19%, this was exceptionally low. Dyeing of the carpet with grease-soluble dyes resulted in no difference in the distribution of the residual grease content.

Further cross-sections in different areas of the carpet were prepared and microscopically examined. Due to differences in the plytwist the loops opened to varying degrees. Under close scrutiny, differences in the plytwist can even be recognized, Fig. 217. Therefore, in the affected areas, the surface was more translucent. More light was able to penetrate through to the backing, thus giving the impression of a color streak.

Fig. 217
Page 142

5.4.4 Streaks and Bars in Cotton Velvet Due to Differences in the Twisting of the Pile Yarns – Practical Example

The pile warps of upholstery velvet were made of a cotton ply yarn. Dyed pieces diplayed a streakiness and barré parallel to the threads which could not be detected in UV light.

No differences in the add-on of lubricating agents could be established between the streaky and the faultless fabric in extraction and dyeing tests with grease-soluble dyes.

Further, film imprints were prepared from the streaky upholstery velvet. Velvet is a problematical fabric because the pile tufts are more or less strongly flattened, so structural unevenness cannot always be identified in this way. During this test the contact pressure was therefore kept as low as possible. Accordingly, streaks on the imprint could be recognized – although only weakly – both in the original and the extracted fabric. Oily and/or greasy residues could not be the causes. The streak formation had to be attributed to structural reasons.

This was substantiated by subsequent microscopic examination of the isolated pile tufts. It was shown that the cut length of the pile tufts and the length of each side were uneven, Fig. 218. Moreover, the test revealed a different untwisting of the pile tufts. Differences in the plytwist of the yarns might be the cause; this could be observed at the base of the tufts.

Fig. 218
Page 144

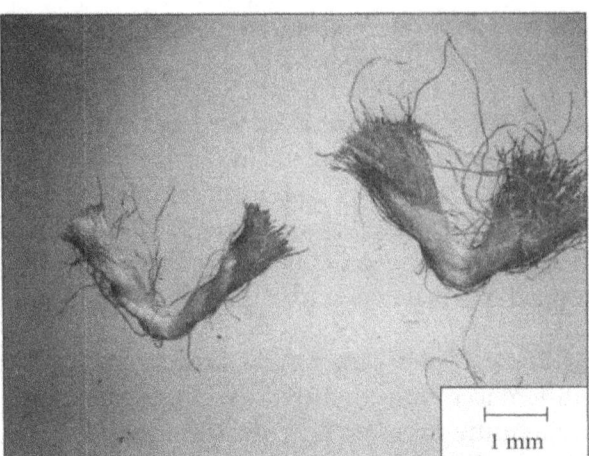

Fig. 218. Pile tufts of a cotton velvet with different twist and different side lengths.

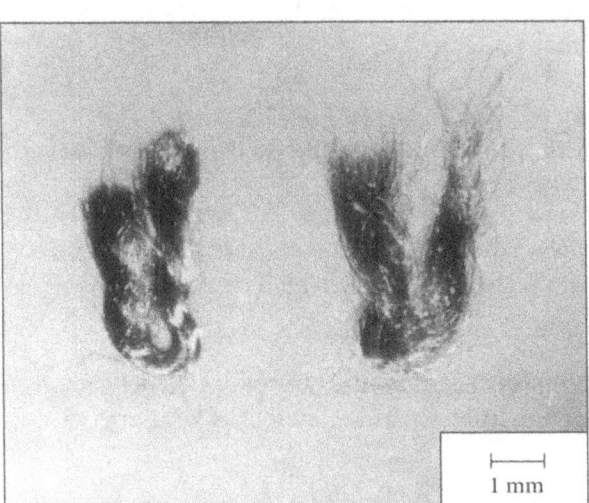

Fig. 219. Tufts from a streaky viscose staple plush with differences in opening and unequal side lengths.

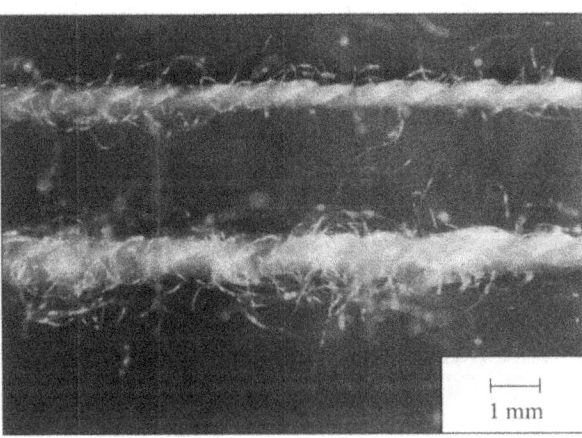

Fig. 220. Yarn for the production of viscose staple plush in Fig. 219 with clear differences in the plytwist.

5.4.5 Streakiness in a Viscose Staple Plush Due to Yarn Differences – Practical Example

A plush made of viscose staple yarn 50 tex × 2, dyed on cross-wound bobbins, displayed a streakiness parallel to the threads. This was first considered to be a dyeing defect.

Pile tufts were isolated from lighter and darker streaks and microscopically examined. Tufts from lighter streaks were more open and voluminous than the other tufts. Additionally, the sides of these tufts were of different length, Fig. 219.

Fig. 219
Page 144

Microscopic examination of the plied yarn utilized for the production of the plush revealed that differences in the volume of the yarns occurred within both shorter and longer thread lengths, Fig. 220, due to differences in the spinning twist and plytwist.

Fig. 220
Page 144

As a result of these yarn differences, the tufts opened to varying degrees during cutting. With more open tufts, incident light was scattered more strongly than with others, causing a lighter coloration.

6 Causes of the Formation of Tight Threads and Their Effects

Tight threads cause a wavy and/or blister-shaped contraction of the fabric [48]. In addition, they can lead to a more or less distinct streak formation. Greater tension of individual weft yarns or warp threads forces the neighboring, slightly stretched threads to be incorporated into the fabric in waves.

Often, tight threads can only be detected when the fabric is wet. The wave form of the fabric can often be eliminated by ironing, but it recurs when the fabric becomes wet again.

Tight threads can have different causes:

a) different moisture content of the yarns during processing,
b) large differences in the twist of the yarns,
c) different degrees of drawing in synthetic fibers.

Several examples will show how tight threads can be recognized as a cause of textile damage.

6.1 Tight Threads in Wool Fabrics Caused by Uneven Yarn Moisture

If wool yarns contain a high and unevenly distributed moisture content, elongation differences occur within the thread during the winding processes. This is especially true in the case of high winding speeds. During drying, the stretched thread sections are fixed in such a way that the tight sections can hardly be noticed in the unfinished material. Only when the fabric is wet will the overstretched thread sections contract to their original length. So-called tight areas are the result.

6.1.1 Tight Picks in Wool Fabrics – Practical Example

Various worsted fabrics from the same manufacturer displayed an uneven, streaky appearance in the weft direction. Strangely, this only occurred during periods of rain.

This defect could be easily recognized both in reflected and transmitted light, Fig. 221 and 222. In the respective areas, more light penetrated over varying lengths of thread. Therefore, it was presumed that different yarn count or twist accounted for this effect.

Fig. 221–222
Page 148

Weft yarns were isolated from both the streaky and the intact areas in the piece and microscopically examined. In the defective areas weft yarns were found to be very tight. In some areas, the weave crimp of the weft yarns could no longer be recognized, Fig. 223.

Fig. 223
Page 148

All rejected fabric samples displayed a marked wavy or blistery appearance after uniform moistening, Fig. 224 and 225, thus proving the existence of tight picks. In this case, yarn with different moisture contents had been processed.

Fig. 224–225
Page 149

6.1.2 Tight Threads in the Warp of a Wool Fabric – Practical Example

A section of a pure wool fabric showed streakiness parallel to the warp within varying lengths of threads. This was blamed on local excessive sizing of the warp.

However, preliminary examination showed that the fabric was properly desized.

Streakiness could also be observed on the film imprint, Fig. 226. Simple dyeing variations would not have been visible on the imprint. Therefore, the streakiness could only have been caused by structural differences.

Fig. 226
Page 149

Further examination of the fabric revealed that the streaks could also be recognized in transmitted light, i.e. the passage of light was greater in the areas in question than it was in the rest of the fabric. Since there were no differences in yarn count and twist, this could only be attributed to the presence of tight threads in the fabric. This assumption was confirmed by the fact that both the dyed fabric and the untreated material, uniformly moistened with distilled water, showed a marked wavy and/or blistery appearance, Fig. 227. This was caused by differences in the shrinkage of warp threads. The fabric samples could be smoothed by ironing; after the next moistening, however, the same material defect recurred.

Fig. 227
Page 149

Fig. 221. Worsted fabric in reflected light with streaks in the weft caused by tight threads.

Fig. 222. Worsted fabric as shown in Fig. 221, but in transmitted light.

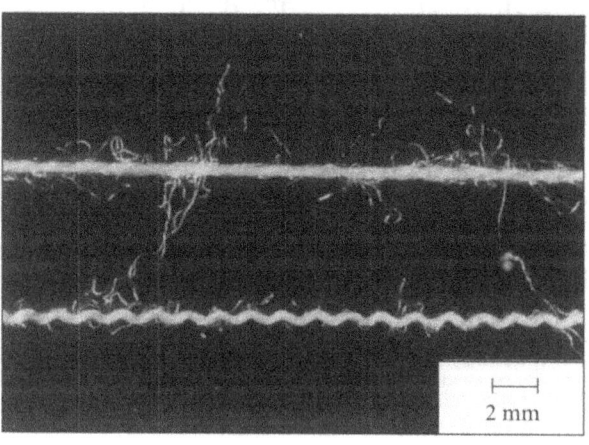

Fig. 223. Weft yarn in normal state (below) and in over-stretched state, from the fabric in Fig. 221.

2 mm

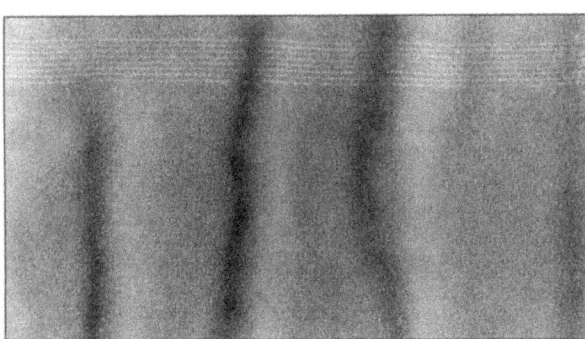

Fig. 224. Wavy appearance of a moistened worsted fabric, caused by tight threads in the weft.

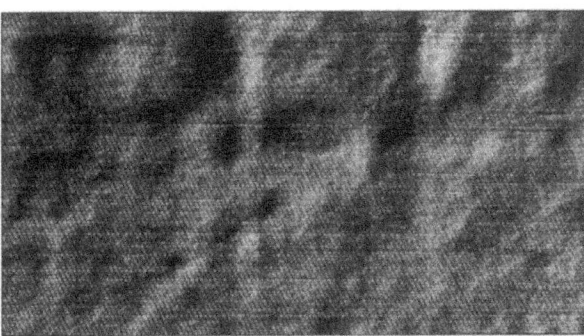

Fig. 225. Blistery appearance of a moistened worsted fabric, caused by tight threads in the weft.

Fig. 226. Film imprint of a streaky wool fabric. Warp streakiness caused by tight threads can be clearly recognized.

Fig. 227. Blistery appearance of the wool fabric in Fig. 226 after moistening, caused by tight threads in the warp.

6.2 Tight Picks in a Fabric Made of Viscose – Practical Example

Tight threads also occur in viscose filament and staple. This is due to the fact that elastic elongation compared to permanent elongation is lower than that of natural fibers, for example. When the yarn is overstretched, it does not return to its original length after relaxing. The overstretched threads try to contract when they become moist; therefore, stretch areas (in the pieces) can be clearly detected in the wet state.

Irregular drawing due to defective braking and guiding equipment frequently causes tight threads. This is especially true if yarns with different moisture contents are processed. Under strain, moist yarn stretches more than dry yarn.

Fig. 228
Page 151

A decorative fabric containing viscose filament in the warp and viscose staple ply yarn in the weft partially contracted into a wavy and honeycomb form when moistened, Fig. 228. The material could be smoothed by ironing. But as soon as the fabric became moist again, the blistery state recurred.

By isolating the weft yarns, differences in length could be detected after removing the weave crimp; the tight threads were clearly shorter, thus proving that there were tight picks.

6.3 Tight threads Caused by Different Yarn Twist – Practical Example

A blue/white checked pillow case occasionally displayed a blistery, crumpled, bark-like crepe appearance.

The blistery, crepe-like state of the pillow case could be smoothed by ironing. But after sprinkling with water the same blistery appearance recurred.

Fig. 229
Page 151

This textile defect was obviously caused by weft yarns because the beginning of the blistery, wavy section ran precisely parallel to the weft, Fig. 229. It was assumed that there might possibly be a mixture error in the weft yarn.

Microscopic examination showed that pure cotton yarns had been processed in both the warp and weft. No encrustations, deposits or chemical damage to the fiber material could be detected. Subsequently, a large film imprint was prepared. On the imprint, weft streaks could be recognized in the wavy area of

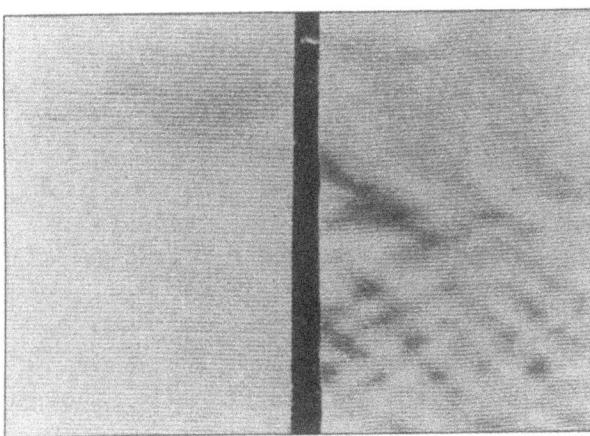

Fig. 228. Fabric made of viscose with tight threads in the weft. Left: Dry, smooth. Right: Moistened, blistery.

Fig. 229. Pillow case with tight threads in the weft after first moistening. Left: More blistery and crumpled, crepe-like state. Right: Smooth fabric area.

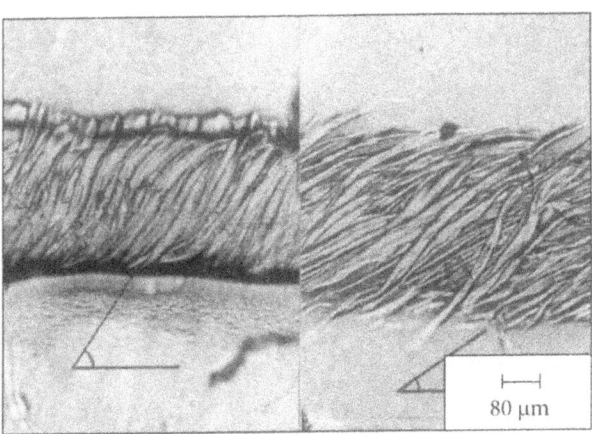

Fig. 230. Film imprints of the weft yarns of the pillow case from Fig. 229. Left: Strongly twisted weft yarn from the crepe-like part of the fabric. Right: Less twisted weft yarn from the smooth part of the fabric.

80 µm

the fabric, but only in the area of the white weft yarns, not the blue. Therefore, there must have been structural differences within the weft yarns.

Fig. 230
Page 151
The weft yarns from the wavy and smooth parts of the fabric were isolated. Film imprints were prepared which were subsequently compared under the microscope. Figure 230 shows that the white weft yarn was twisted more strongly in the wavy area than in faultless parts of the fabric. The different twisting of the yarns can be easily recognized on the imprint, taking into consideration the angle between the fibers and the thread axis.

The crepe-like and/or blistery appearance of the fabric is a result of different twist level of the yarn used in the weft. The very strongly twisted white weft yarn looked like a crepe thread and contracted in the wet state.

7 Defects Caused by Deposits and Encrustations on the Fiber Material

Oils, greases and waxes (e.g. from spin finishes, lubricants, coning oils, sizing waxes and loom oils) which are not removed before dyeing can cause reserving and, as with sizing residues, lead to dyeing unlevelness. Precipitated dye or undissolved dye particles cause dye stains. Inappropriate finishes lead to the formation of chalky marks when the fabric is scratched. Oligomer and lime deposits result in graying and light stains on dyed and printed fabric.

7.1 Detection of Oil, Grease, Paraffin or Wax Deposits by Means of Dyeing with Oil-Soluble Dyes

Oils, greases and waxes can be detected by means of dyeing with oil-soluble dyes. The dyeing process is described in detail for one well suited dye, Sudan Red 460 (BASF); other dyes, e.g. Fat Red 5B (Hoechst), can be used if the respective processing conditions are taken into consideration.

0.5–1 g/l of Sudan Red 460 are made into a paste with 10 ml of methanol by stirring for 3 minutes. The pasted dye is then doused with 1 liter of water at 40 °C, and 10 ml of conc. hydrochloric acid is added. The dye dispersion is not filtered.

The fabric sample is treated in the dye dispersion for approx. 10 minutes at 40 °C. If necessary, the temperature can be increased to 70 °C, especially if there are deposits of paraffin, high-melting waxes etc. The sample is then rinsed for 3 minutes under running water and dried at 100 °C. In the presence of oils, greases, waxes or paraffin the textile is dyed intensively red, Fig. 231.

Fig. 231
Page 154

For microscopic examination, the fiber material is immersed in a dye solution of 0.3–0.5 g of Sudan Red 460 (dissolved in 50 ml of ethanol, with subsequent addition of 50 ml of glycerol). Dyeing of the greasy or oily areas can be observed immediately after immersing.

Practical examples shall illustrate grease detection with this dye.

Fig. 231. Detection of oil and/or grease soiling with a red oil-soluble dye.

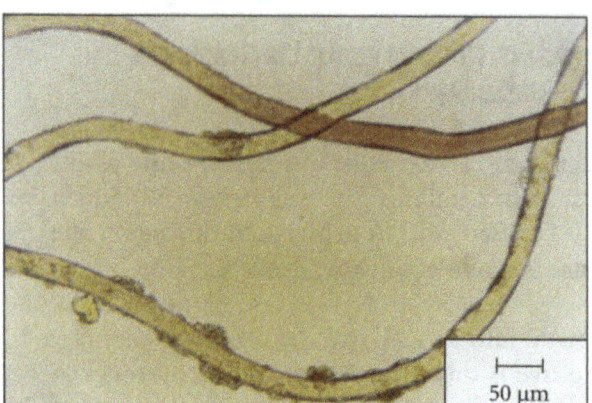

Fig. 232. Polyamide fibers, reserved during dyeing due to oil or grease deposits.

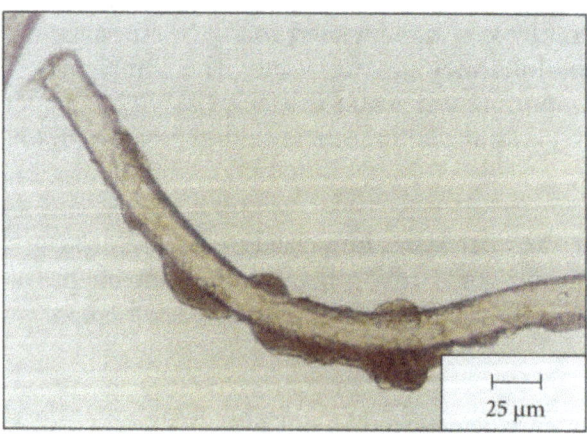

Fig. 233. The same poly-amide fibers as in Fig. 232. Deposits dyed with Sudan Red 460.

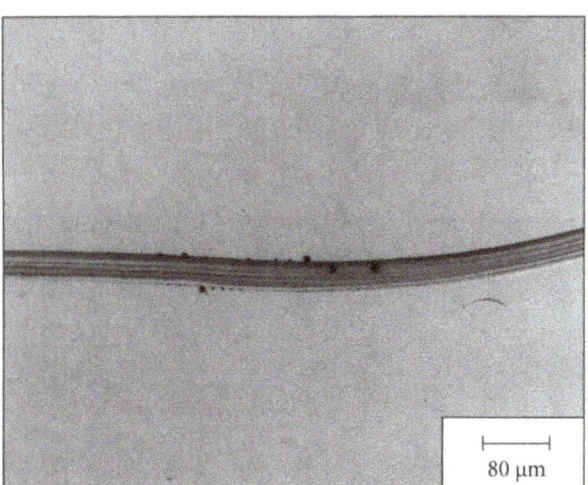

Fig. 234. Viscose fiber with spin finish residues, dyed under the microscope with the reagent Sudan Red/glycerol/alcohol.

Fig. 235. Dyeing of a raw wool fabric with Sudan Red 460. Strong streak formation in weft direction, caused by differences in the lubricant pick-up.

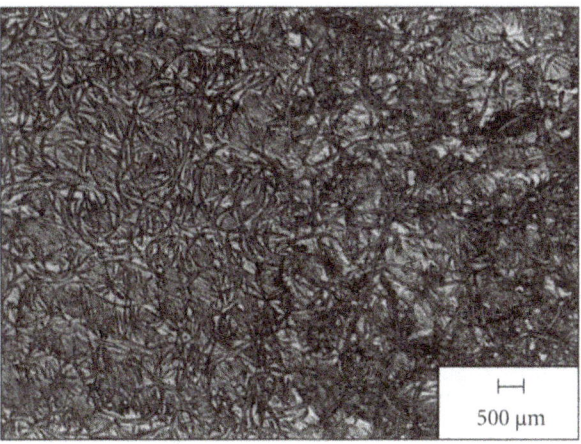

Fig. 236. Film imprint of the fulled cloth of the wool fabric in Fig. 235. Differences in the lubricant pick-up lead to differences in the felting of the surface.

7.1.1 Dye Resisting Effects of a Polyamide Fabric – Practical Example

A knitted fabric made of textured polyamide showed light stains after dyeing. This damage could not be eliminated by stripping and redyeing of the pieces, the light stains remained unchanged.

Fig. 232
Page 154 Microscopic examination revealed deposits in the areas of the light stains, Fig. 232; thus the fibers were not dyed in these areas.

Fig. 233
Page 154 The oil-soluble dye Sudan Red 460 dyed the deposits red, Fig. 233, thus proving that the deposits were grease soiling.

7.1.2 Warp Streakiness in a Lining Material
Caused by Failure to Adequately Wash out the Spin Finish
– Practical Example

A viscose lining material showed streaks parallel to the warp. The streaks were 1 mm wide and 30–40 mm long. In order to determine the cause of this material defect, the lighter threads in the warp were microscopically compared to normally dyed warp threads. In this case, encrustations due to sizing residues could not be found.

Fig. 234
Page 155 Shortly after immersion of the defective fiber material in Sudan Red/glycerol/alcohol, small globules dyed intensively red could be recognized on the surface of the fibers. These phenomena, which are typical of oil and/or spin finish residues, only occurred on the lighter dyed warp threads, Fig. 234. The reserving was thus caused by oils which were difficult to wash out. After stripping and thorough scouring, this fabric could be dyed without further problem.

7.1.3 Streaks in Fulled Wool Fabrics Due to Differences in the Fiber
Lubricant Pick-Up – Practical Example

Even if a lubricant, e.g. coning oil, can be washed off easily, problems can arise during fulling if variations in the lubricant pick-up within one piece are too great. In the case of yarns with a strongly overdosed lubricant pick-up, it takes slightly longer until the oil is removed and the actual felting process in the fulling machine starts. In this case, the lubricant must be washed out before fulling in order to achieve uniform felting of the fabric.

A wool fabric showed weft streakiness after dyeing and finishing, causing an uneven appearance of the fabric. It was assumed that it was caused by defective lubrication of the weft yarns. The tops of the cones had been painted with a coning oil before weaving because only this procedure permitted the single yarn to be woven correctly.

The residual grease content of the weft yarns after lubrication was 0.08 – 0.12 %, i. e. the differences were not very large.

In the untreated material, however, great differences in the lubricant pick-up with values between 1.10 to 11.70 % were found. This was confirmed by means of the staining test using Sudan Red 460, Fig. 235.

Fig. 235
Page 155

Fabric imprints were produced of the rejected finished fabric and subsequently examined microscopically. In the vicinity of the rejected, slightly darker streaks, the threads were not as open, voluminous and felted as in the rest of the fabric, Fig. 236.

Fig. 236
Page 155

Fiber damage could be excluded as a cause for the defective appearance because the wool fibers had a distinct scale structure with no sign of damage, Fig. 237.

Fig. 237
Page 158

7.1.4 Detection of Oil and/or Grease Soiling on Polyester

In the finishing process, removal of oil and/or grease soiling from a polyester fabric is often problematic; the lipophilic character of these synthetic fibers plays an important role. The polyester fiber absorbs oils and greases. The longer fabrics with oil and/or grease soiling are stored, the more the oils and greases penetrate into the fibers. Heat, e.g. generated through singeing or setting, encourages diffusion into the fiber interior. This makes it extremely difficult to wash out the soiling. In extreme cases, traces of penetrated oil or grease can act as carriers during dispersion dyeing. In addition, burnt-in, greasy substances are very difficult to detect.

Figure 238 shows a dyed polyester fabric with a dark grease stain; Fig. 239 shows grease detection with Sudan Red 460, and Fig. 240 a single fiber isolated from the dyed stain. Apparently the grease has already diffused into the fiber to a large extent.

Fig. 238-240
Page 158-159

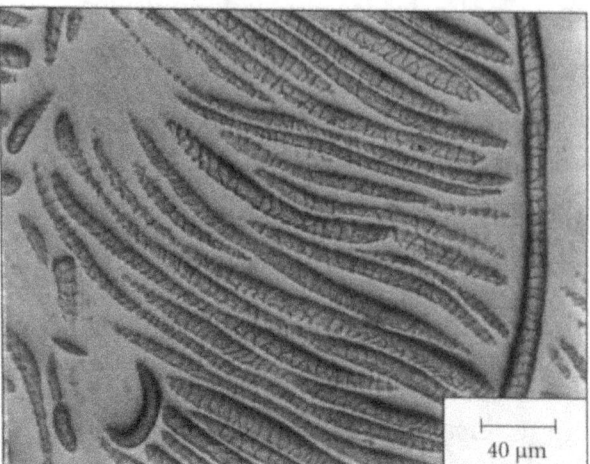

Fig. 237. Magnification of the film imprint in Fig. 236. The individual wool fibers in the streaky areas are intact.

Fig. 238. Dyed polyester fabric with dark oil stain.

Fig. 239. The same fabric as in Fig. 238, dyed with Sudan Red 460. Oil detection is relatively weak, since soiling has diffused into the fibers during drying and setting.

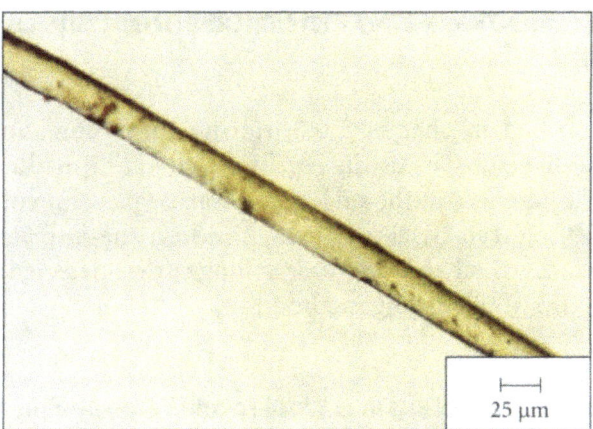

Fig. 240. Isolated fibers from the stained fabric in Fig. 239, dyed with Sudan Red 460.

Fig. 241. Film imprint of a wool fabric with lighter dyed areas. Dye reserving is caused by a loom oil which is difficult to wash out.

Fig. 242. Film imprint of a cotton fabric with reserved areas after dyeing. The oil soiling and marking lines can be clearly recognized on the imprint.

7.2 Detection of Oil, Grease, Wax and Paraffin Deposits by Means of Film Imprints

As has already been described in chapter 1.6.2, during the preparation of imprints, oily and grease-like substances diffuse into the softened film and make it dull. If there are differences in the oil, grease, wax or paraffin content within the textile fabric, these can be clearly recognized on the imprint. By means of film imprints, traces of oil and/or grease soiling can be detected even more easily than with the aid of the dyeing method [49].

7.2.1 Light Stains Caused by Oil Soiling in a Wool Fabric After Dyeing – Practical Example

A dark blue wool fabric showed slightly lighter stains after dyeing. It had to be determined whether it was a dyeing defect or whether deposits had caused reserved areas during dyeing.

Fig. 241
Page 159 For this purpose, a large film imprint was taken of the stained fabric. As shown in Fig. 241, the stains were also clearly reproduced on the imprint. Thus it could not be a dyeing defect because this would not be visible on the imprint.

After extraction with petroleum ether, the lighter stain was still recognizable on the fabric. On an imprint of this fabric, the stains were no longer recognizable. Thus oil and/or grease soiling must have caused reserving during dyeing. This was confirmed by a dyeing test with Sudan Red 460 which was slightly positive in the reserved areas.

A check in the plant showed that the stains had been caused by a loom oil which is hard to wash out.

7.2.2 Reserved Areas Due to Oil Soiling in a Cotton Fabric – Practical Example

After dyeing, a cotton fabric showed light stains parallel to the warp. These areas had obviously been reserved during dyeing. Under UVlight they showed brighter fluorescence.

Fig. 242
Page 159 On a film imprint the stains were clearly reproduced in the form of dulled areas, Fig. 242. Apart from this, marking lines can be easily recognized which

had been written on the fabric with a ball-point pen in order to identify the defects. This example illustrates the precision of the imprint technique for the reproduction of deposits.

After extraction of the stained fabric with petroleum ether, no more light areas could be found on the imprint. Thus it can be concluded that it was oil and/or grease soiling. This conclusion was confirmed by dyeing tests with oil-soluble dyes, although the results were not as clear as the dulling on the polystyrene film.

7.2.3 Streak Formation in Knitwear Caused by Uneven Paraffination – Practical Example

Paraffination of yarns for ease of processing on knitting and/or hosiery machines is problematic for the finisher, particularly if it is uneven. Homogeneity of paraffination cannot always be determined by dyeing tests with oil-soluble dyes, since natural fats and waxes of cotton are also dyed.

On film imprints, the natural waxes of cotton do not interfere since they are distributed uniformly. Therefore, even small differences in paraffin add-on can be detected by means of film imprints. This is explained in detail by the following practical example.

A piece of knitted fabric from a yarn-dyed blend of acrylic and cotton showed darker streaks over shorter thread lengths. After petroleum ether extraction of a fabric sample, the streaks were no longer visible. For the extractable proportion a value of 0.94 % resulted; according to the IR spectrum it mainly consisted of hydrocarbons.

Next, a film imprint was produced of the extracted and the non-extracted fabric, Fig. 243. The differences in paraffination of the yarns were clearly visible on the imprint of the non-extracted sample, while there were no streaks on the imprint of the extracted, paraffin-free fabric. This proved that the dark streaks were caused by uneven paraffination of the yarn.

Fig. 243
Page 162

7.3 Detection of Pigment Deposits on Imprints

Loose pigment deposits on the fabric, e.g. resin particles, dye, lime and oligomers, adhere to the soft film and thus become visible on the imprints. In this

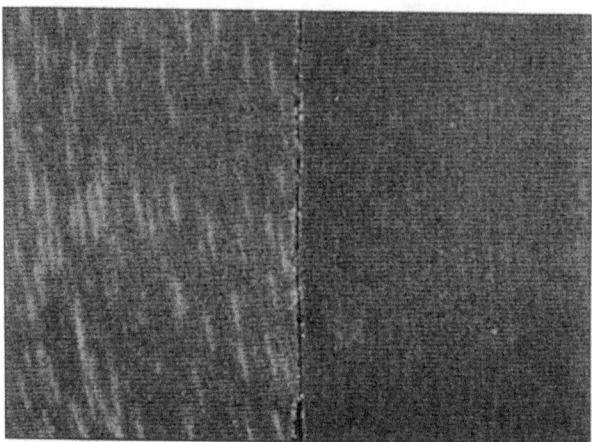

Fig. 243. Film imprint of a knitwear blend made of acrylic/cotton with different paraffination. Left: As delivered. Right: After extraction.

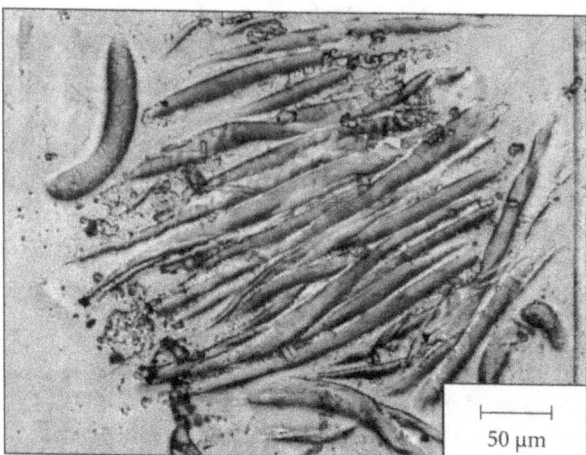

Fig. 244. Film imprint of a stain on a fabric made of polyester/cotton. Resin deposits can be recognized.

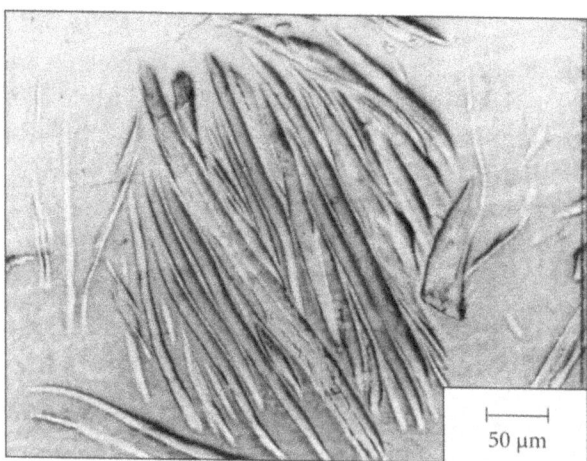

Fig. 245. Film imprint of an intact area of the fabric in Fig. 244. The fibers are free of encrustations.

way the pigments on the films can be examined microscopically. Separation of textile fabrics into threads and individual fibers can thus be avoided, which will be explained by means of examples.

7.3.1 Reserved Areas in a Polyester/Cotton Fabric Due to Resin Deposits – Practical Example

A fabric made of polyester/cotton was dyed and finished with pigment dyes in a continuous process. A mixture of synthetic resin, catalyst, pigment printing binder and pigment dye was used as a finishing liquor. After drying and fixation there were light stains on the fabric which were assumed to have been caused by addition of a silicone defoamer to the finishing liquor.

In order to determine the cause, large film imprints of the stained fabric were produced which also showed the stains.

Further examination of the fabric revealed that the white stains were not dyed by oil-soluble dyes. Thus oils, greases and waxes could not have caused the stains. The silicone foam inhibitor used in this case could also be excluded since it was weakly dyed by the oil-soluble dyes.

For microscopic examination, one can use the large film imprint and cut it into small pieces. If, however, the large imprint is to be kept for evidence, it is recommended to produce film imprints of the stained areas with the size of microscope slides.

Examination showed that there were resin encrustations in the stained areas, Fig. 244 and 245. Apparently the synthetic resin had partially hardened in the finishing liquor, thus causing resin deposits and stain formation. Fig. 244 – 245 Page 162

7.3.2 Stain Formation Caused by Lime Deposits on Polyester Knitwear – Practical Example

After finishing, a dark blue dyed polyester fabric contained grayed areas. These were allegedly caused by aftertreatment with a cationic softener.

Extraction with petroleum ether did not change the light and/or grayed areas. A film imprint was therefore taken of a stained area. Microscopic examination showed deposits with a crystalline structure, Fig. 246. These deposits were relatively easy to remove mechanically from the fabric. Fig. 246 Page 165

Fig. 247
Page 165

The fabric sample was then ashed and dissolved in 2 N hydrochloric acid p. a. One drop of this hydrochloric solution was subsequently placed on a microscope slide. Directly next to it 2 N sulphuric acid p.a. was dripped onto the slide. The microscope slide was covered with a cover glass. Under the microscope, formation of calcium sulfate crystals could be observed, Fig. 247. Thus the gray stains were lime deposits which could be removed by acidification of the fabric with hydrochloric acid.

7.3.3 Pigment Soiling on a Plyed Yarn Made of Acrylic Fibers – Practical Example

An awning fabric dyed in the flock showed dark streaks over short thread lengths. Since these streaks could only be found on the warp threads, it was assumed that they had developed during sizing.

Fig. 248
Page 165

For verification, streaky warp threads were isolated and imprinted onto films. Microscopic examination of the imprints showed that it was pigment-like soiling, part of which stuck to the film. Only one thread in the plyed yarn had been soiled, Fig. 248. Since it is extremely improbable that during sizing only one particular thread in the ply could be soiled, the material must have been soiled before or during twisting. Thus the sizing plant could not be blamed for this production defect.

7.3.4 Speck-Like Dark Stains on a Cotton Fabric Caused by Undissolved Dye Particles – Practical Example

Fig. 249
Page 166

A pure cotton fabric dyed light gray showed numerous small, speck-like dark stains. In order to find the cause of this speck formation, imprints were taken on polypropylene films which were subsequently examined under the microscope. Here it was observed that occasionally there were dark blue deposits on the stains which stuck to the film and obviously consisted of dye which had either precipitated or was not dissolved, Fig. 249. In the case of printed fabric, non-fixed dye also adheres to the film.

7.3.5 Graying and Light Stains on a Polyester Fabric Due to Oligomer Deposits – Practical Example

On a color-woven parka fabric made of polyester, graying and light stains were visible after scouring. The light stains were also clearly recognizable on a large

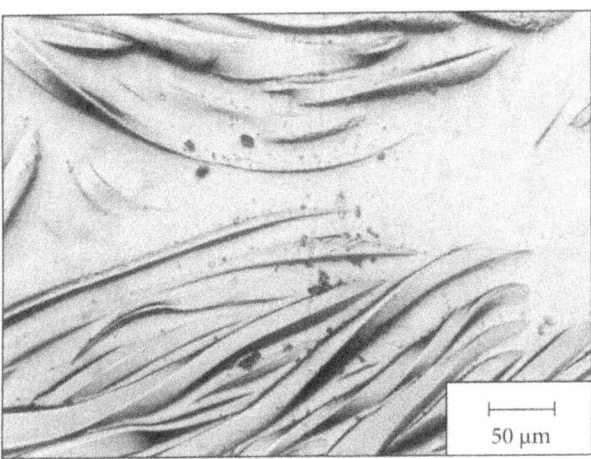

Fig. 246. Film imprint of a fabric made of textured polyester with lime deposits.

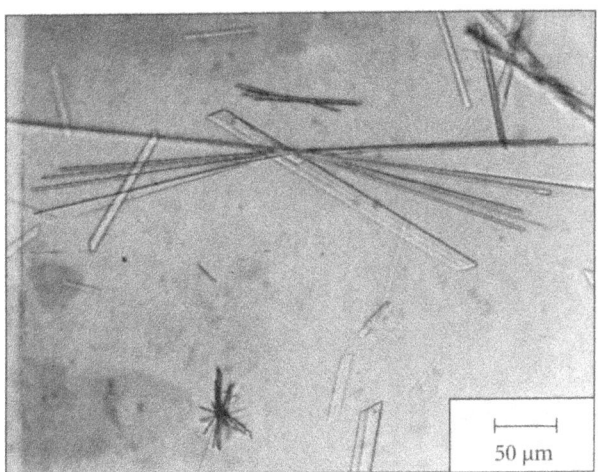

Fig. 247. Microscopic detection of calcium due to formation of calcium sulfate crystals.

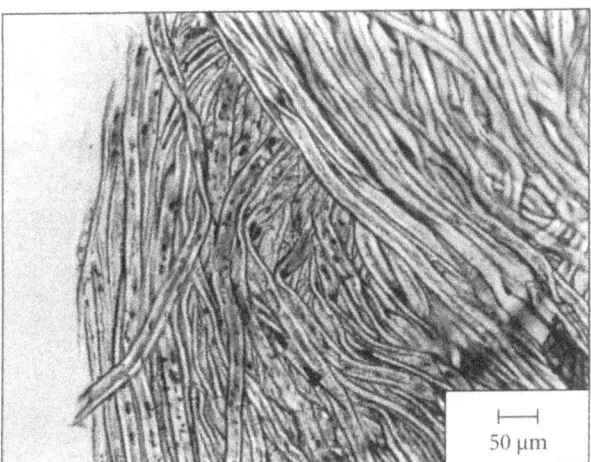

Fig. 248. Film imprint of a yarn made of acrylic fibers with a soiled yarn area. One thread in the ply is clean (top right), one thread is soiled (arrow).

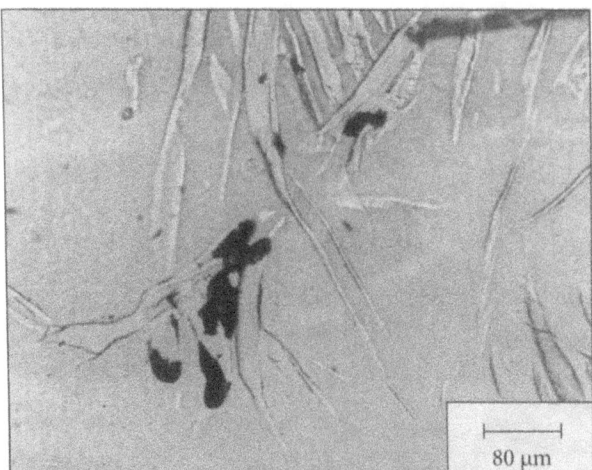

Fig. 249. Film imprint of a cotton fabric with dark stains after dyeing. The precipitated dye adheres to the film.

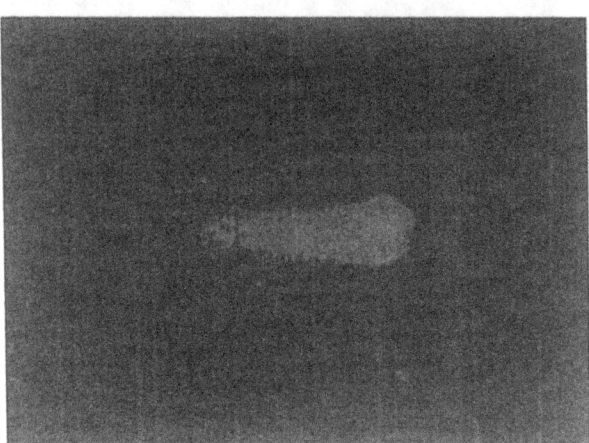

Fig. 250. Film imprint of a polyester fabric with stains and streaks parallel to the weft or warp.

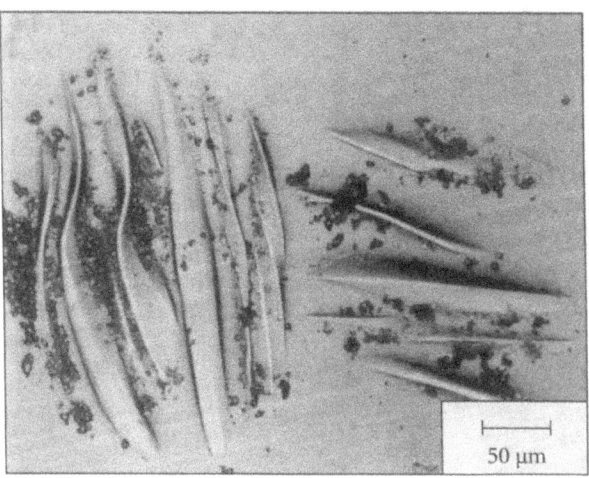

Fig. 251. Film imprint of the stained polyester fabric in Fig. 250, strongly magnified. There are oligomer deposits on the polyester fibers.

film imprint, Fig. 250. The streaks parallel to warp or weft, which were visible on the imprint but not on the fabric because of the stripe pattern, could be identified as yarn differences caused by texturing differences.

Fig. 250
Page 166

Microscopic examination of the film imprint clearly revealed that the stains were deposits, Fig. 251.

Fig. 251
Page 166

The deposits were relatively easy to remove mechanically from the fabric. Under the microscope in polarized light they showed birefringence. They had a crystalline structure and a melting point of 317 to 320 °C. After cooling of the sample, the typical crystals of the cyclic trimer of polyethylene terephthalate formed under the microscope, Fig. 252.

Fig. 252
Page 168

7.4 Detection of Film-Forming Products and Film-Like Deposits by Means of Imprints

Film-forming products, e.g. sizing agents and finishes, can also be recognized with the aid of the replication method. This method is especially useful here because it avoids the separation of textile fabrics into threads and fibers required for the preparation of fiber samples (this destroys the films and can lead to misinterpretation).

Figure 253 shows a fabric with a cotton warp and a textured polyester yarn in the weft. The sizing agent on the cotton warp can be clearly recognized as a film-like coating. It can further be seen that the sizing agent does not form a closed film around the warp threads but is partially like a flat cake. Figure 254 shows the same fabric after desizing. The cotton warp threads are more open, more voluminous and free of encrustations and bonds.

Fig. 253 – 254
Page 168

7.4.1 Hardening of the Hand Due to Residues of Printing Paste Thickeners – Practical Example

A printed cotton fabric partially showed hardening of the hand in the printed areas. On large film imprints the affected printed pattern areas could be clearly recognized; thus there had to be deposits. This was confirmed by microscopic examination of the film imprints, Fig. 255, in which a film-like coating was clearly detected in the printed areas. It was a printing paste thickener which had not been scoured off completely.

Fig. 255
Page 170

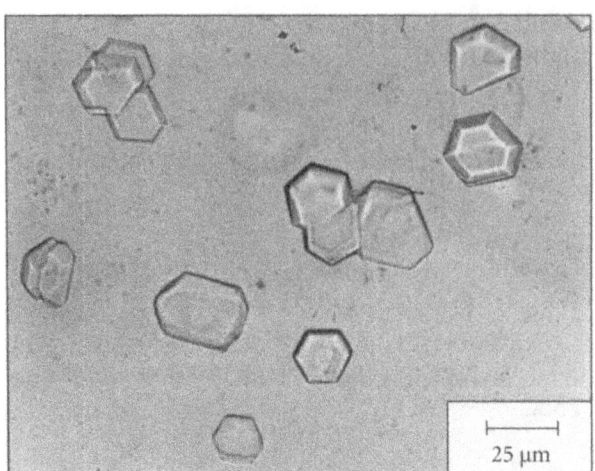

Fig. 252. Crystals formed after melting of the oligomers as a result of recrystallization.

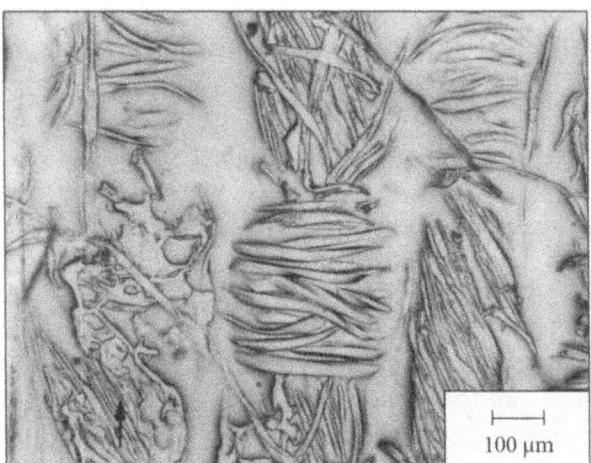

Fig. 253. Film imprint of a fabric made of polyester/cotton. Weft: Textured polyester yarn. Warp: Sized cotton yarn.

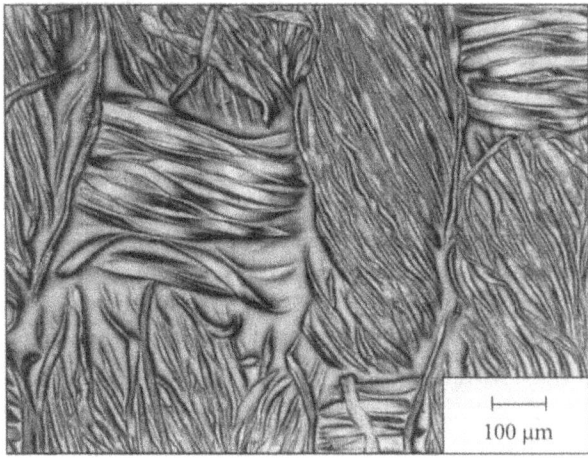

Fig. 254. The same fabric as in Fig. 253, but in the desized state.

7.4.2 Wool Fabric Showing Chalky Marks when Scratched, Caused by the Backing – Practical Example

Wool fabrics which had been finished on one side with polyvinyl acetate foam were rejected because they showed chalky marks when scratched on the upper side of the fabric.

In order to find the cause, film imprints were taken of the surfaces of the areas with and without this fault from the right fabric side and compared microscopically. On the imprint of the faultless area, no finish substance could be detected, Fig. 256. On the imprint of the faulty area, the finish was in some parts clearly detectable, Fig. 257. Obviously there had been an error during application, which led to the penetration of the finish and thus to the marking phenomenon.

Fig. 256 – 257
Page 170

7.4.3 Printed Fabric Made of Silk/Viscose with Hardened Areas Caused by Adhesives for the Printing Table – Practical Example

A printed jersey fabric made of silk/viscose showed grayed and hardened areas which could not be removed by after-scouring at 40 °C.

Microscopic examination of film imprints of this fabric showed that at the hardened areas the fiber material had been bonded with a film-like substance, Fig. 258. Under the microscope it could be recognized as a clear transparent film. In staining tests with iodine/dioxane/boric acid it was identified as polyvinyl alcohol. The mentioned reagent is produced as follows:

Fig. 258
Page 171

20 ml N/10 iodine solution,
80 ml dioxane,
 1 g   boric acid.

The boric acid is dissolved by heating, the solution cooled down to 25 °C, and dioxane is added to obtain 100 ml. Then 7 ml of dist. water are added. Finally the solution is shaken thoroughly. To detect the polyvinyl alcohol, 2 – 3 drops are dripped onto the test fabric. After 20 minutes the polyvinyl alcohol is dyed dark blue to bluish green.

The film had formed because the printing table adhesive layer applied was much too thick and the scouring temperature of 40 °C was not sufficient to remove it from the fiber material. The film could only be removed in a post-

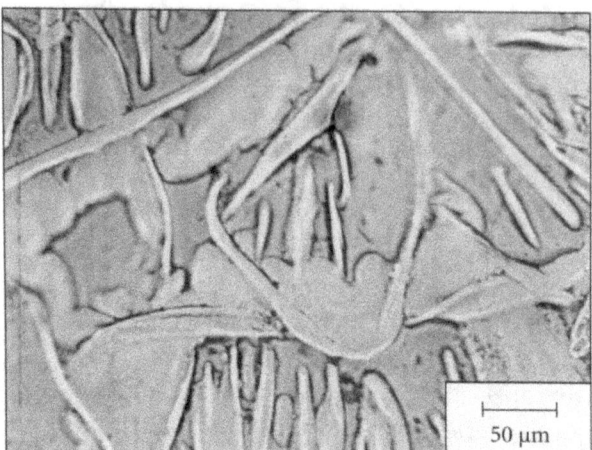

Fig. 255. Film imprint of a printed cotton fabric with a film-like coating of thickener residues.

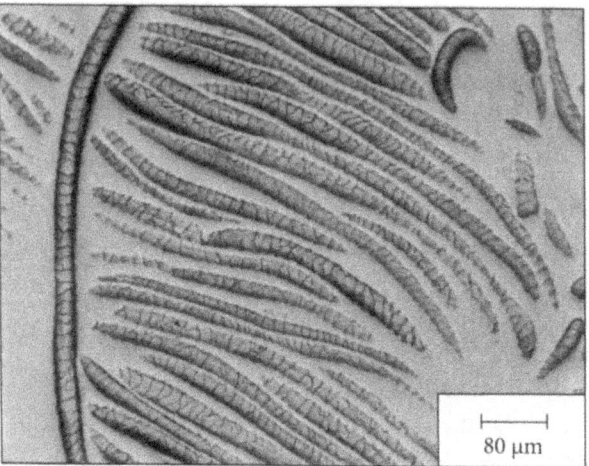

Fig. 256. Film imprint of the upper side of a wool fabric with backing: clean, no penetrated finish.

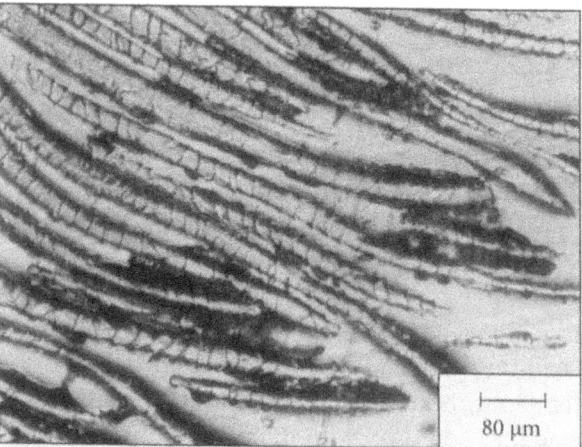

Fig. 257. Film imprint of an area showing chalky marks when scratched, from the wool fabric from Fig. 256. The backing has penetrated through to the upper side of the fabric, the finish particles are clearly visible.

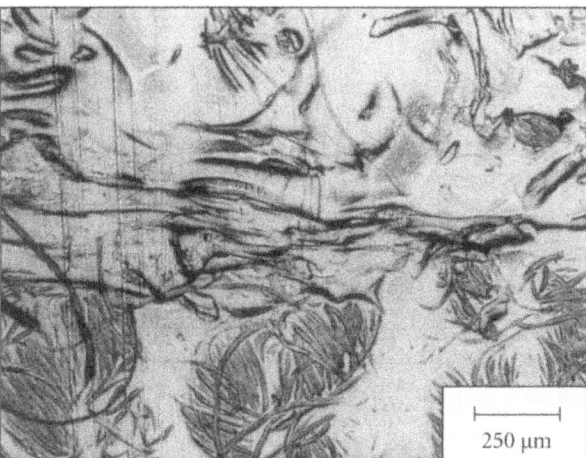

Fig. 258. Film imprint of a hardened area in a silk fabric; a film-like coating is recognizable, caused by the printing table adhesive used.

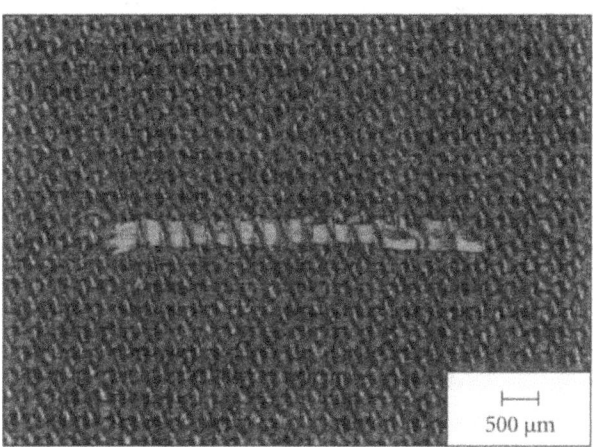

Fig. 259. Fabric made of polyester/cotton with light impurities in the weft.

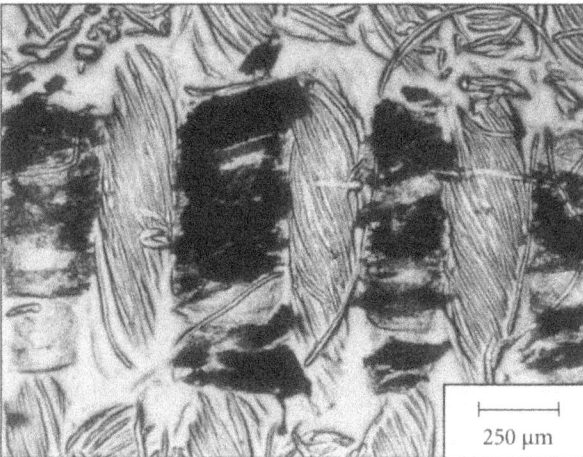

Fig. 260. Magnified film imprint of the same fabric as in Fig. 259 with thermally damaged polyvinyl alcohol sizing agent which adheres to the film.

scouring process at 80 °C; however, the green background dye bled slightly, so that the shade changed towards blue-green.

7.4.4 Deposits of Sizing Agent on the Weft Yarns of a Polyester/Cotton Fabric – Practical Example

Fig. 259
Page 171

A fabric made of polyester/cotton was rejected because of small film-like deposits which were partially incorporated in the fabric and appeared as light areas after dyeing, Fig. 259. The film-like deposits in part had a brownish color. However, this fabric damage was only found on material which had been produced on broad and high-speed shuttle looms.

Fig. 260
Page 171

A film imprint was taken of a defective fabric section and examined microscopically. The examination showed that the film-like deposits could partially be imprinted on the film, or they partially stuck to the film. These particles were of a more or less brownish color, which can obviously be attributed to stronger thermal action, Fig. 260.

Microscopic examination of the film-like deposits in staining tests (iodine dioxane boric acid, chapter 7.4.3) showed that it was a polyvinyl alcohol sizing agent. The friction of the shuttle must have caused the shuttle and the warp material to heat up strongly. Thus the polyvinyl alcohol sizing film was thermally damaged at the surface; it became brittle, peeled off and caused the rejected inclusions which could not even be removed with boiling water.

A check in the weaving mill later revealed that during operation some shuttles developed slight grooves and cracks at the lower sliding surface on which the sizing agent had deposited during weaving. It stuck to the sides of the shuttle or formed deposits in the shuttle box. From there, it was partially transported back onto the fabric, where it was deposited and finally woven into the fabric.

7.5 Detection of Deposits in Staining Tests, Yarn Cross-Sections and/or Fabric Cross-Sections

With negative imprints, deposits can be identified easily, but they cannot always replace the fiber- and product-specific staining tests. Production of yarn and fabric cross-sections – often in connection with color reactions – is sometimes unavoidable in order to find the cause of the damage; this will be explained by means of several practical examples.

7.5.1 Evaluation of Sizing Agent Distribution on Yarn Cross-Sections by Staining of the Starch Sizing Agent with Iodine Solution

Two samples of warp material made of pure cotton, which had been sized with different starch products, were to be evaluated with respect to surface sizing and thread compactness.

For this purpose, after sizing the warp threads were embedded in a polyester casting resin, and microtome sections were prepared. Before microscopic examination, they were immersed in an iodine solution for a short time, in order to make the starch sizing agent visible by means of the blue coloration.

Figure 261 shows a section through a well sized warp thread. The fibers are well bonded, even in the core; the film coating on the thread, the surface size, can be clearly recognized. Figure 262 shows a less well sized thread. The fibers are insufficiently bonded and the sizing film is non-contiguous and torn.

Fig. 261-262
Page 174

In this context it should be emphasized that it is very difficult to evaluate the distribution of the sizing agent on 10 µm thin sections. One must prepare several samples of different threads in order to obtain a general overview.

7.5.2 Oversized Warp Threads – Practical Example

A cotton warp which had been sized with a starch derivative showed bonded areas over the whole width.

For closer examination of the cause, microtome sections were produced of the warp threads, and the starch sizing agent was stained with iodine solution. Microscopic examination showed that the unbonded warp threads were relatively uniformly coated with the sizing agent, Fig. 263. On the bonded warp threads the thicker layer of sizing agent can be recognized clearly, i.e. in these areas the warp was oversized, Fig. 264.

Fig. 263-264
Page 175

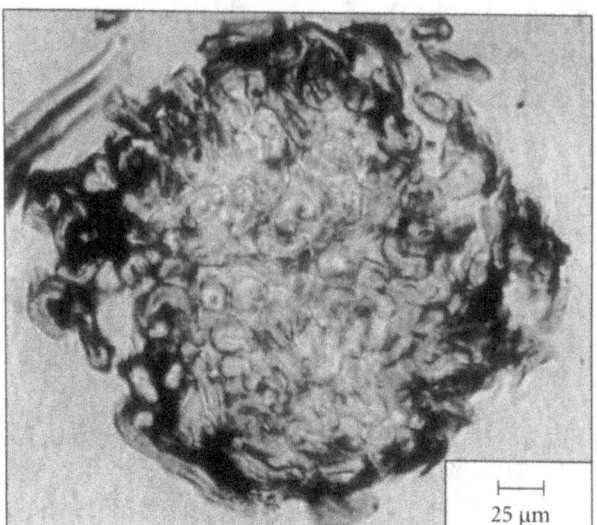

Fig. 261. Cross-section of a cotton warp, starch sizing agent stained with iodine solution. The thread is well coated with the sizing agent.

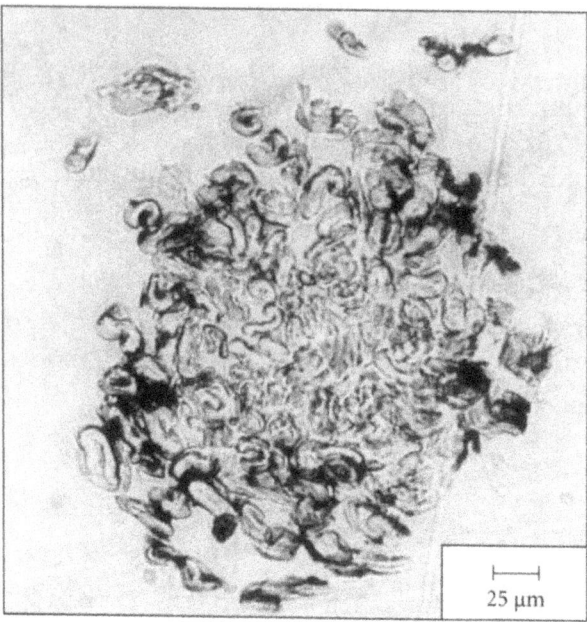

Fig. 262. The same warp material as in Fig. 261. The coating of the thread with the sizing agent is less distinct.

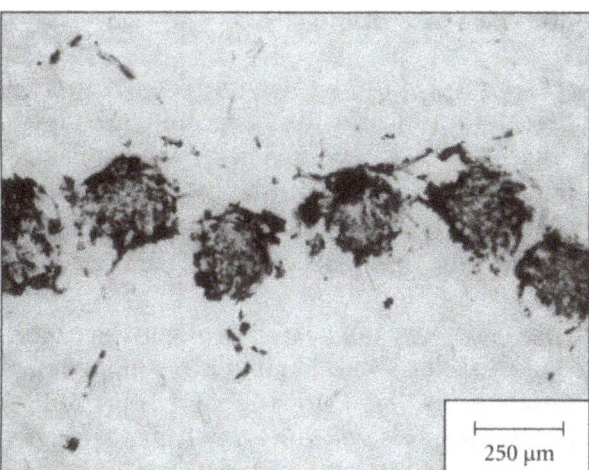

Fig. 263. Cross-section of unbonded warp threads, starch sizing agent stained with iodine solution. The uniform coating of the warp threads can be clearly recognized.

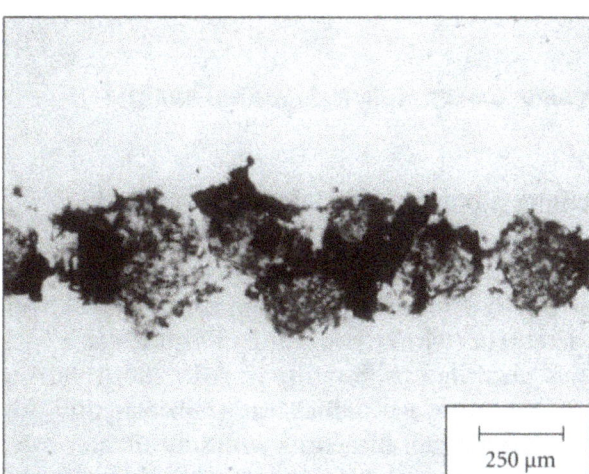

Fig. 264. The same warp material as in Fig. 263. The warp threads are bonded by oversizing of the warp.

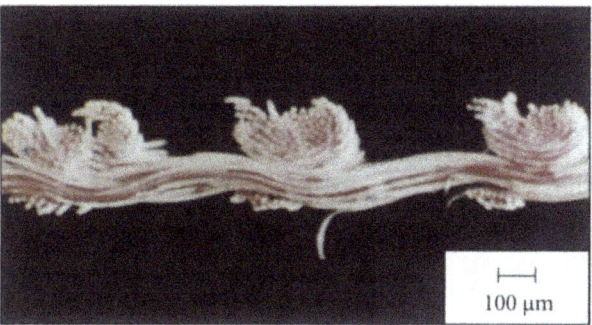

Fig. 265. Cross-section of a polyamide knitwear, durable antistatic agent stained with Sirius Pink BB 143%.

7.5.3 Detection of Durable Antistatics by Staining

Fig. 265
Page 175
Durable antistatics like Nonax 1166 (COGNIS), based on alkaline cross-linkable polyglycol polyamine condensates – similar to sizing agents – do not form an ideal film on the fiber material. The film is deposited in the form of flat cakes on the fibers, Fig. 265. For examination, the durable antistatic was stained with Sirius Pink BB 143%, see chapter 2.4.1, Fig. 84. Due to the intensive red dyeing, the distribution of the antistatic finish on the fiber material can be easily recognized.

The durable antistatics mentioned above can also be dyed with Neocarmin W. The textile material is dyed in dye solution for 5 minutes at room temperature and then rinsed with cold water until the liquor is clear. Neocarmin W dyes these products blue [50]. However, this is not true for polyamide fibers with a permanent antistatic finish. Neocarmin W dyes the polyamide fibers yellow; depending on the thickness of the antistatic layer, this results in a more or less distinct olive green color.

7.5.4 Black Specks in a Polyester Curtain Fabric After Bleaching – Practical Example

Fig. 266
Page 177
After scouring and bleaching, a polyester curtain fabric showed small, black specks, Fig. 266.

The fabric was pre-scoured at 80 °C in an acid medium on a beam dyeing system with a non-ionic detergent (with the addition of 3 g/l oxalic acid + 6 ml of conc. hydrochloric acid) and then thoroughly rinsed. After scouring, the fabric was in a faultless state. The pieces which were subsequently optically brightened also showed no defect. After bleaching with sodium chlorite, however, for which a chlorite-resistant optical brightener was added, small black specks could be recognized on the fabric on the inner layers of the beam when the liquor circulated from the inside towards the outside. With reversed liquor circulation, the specks occurred on the external layers. At first it was assumed that the black specks were particles of marking ink with which the pieces had been numbered.

Fig. 267
Page 177
Figure 267 shows individual fibers from a black speck. In addition to individual, pigment-like deposits there are also areas in which the fibers are covered with a black film-like substance.

Fig. 268
Page 177
Figure 268 shows polyester fibers with marking ink which consists of black pigments and a colorless bonding film. Model tests have shown that this marking

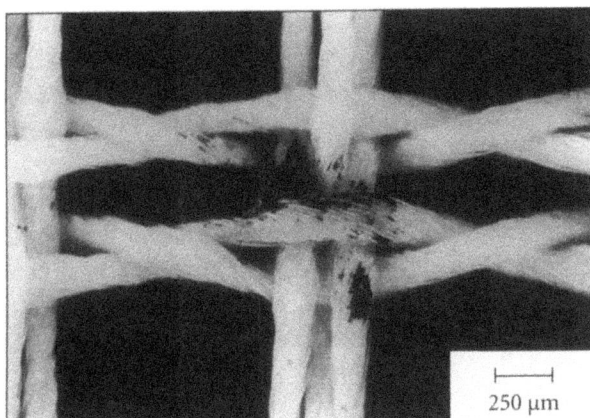

Fig. 266. Polyester curtain fabric with small, black specks.

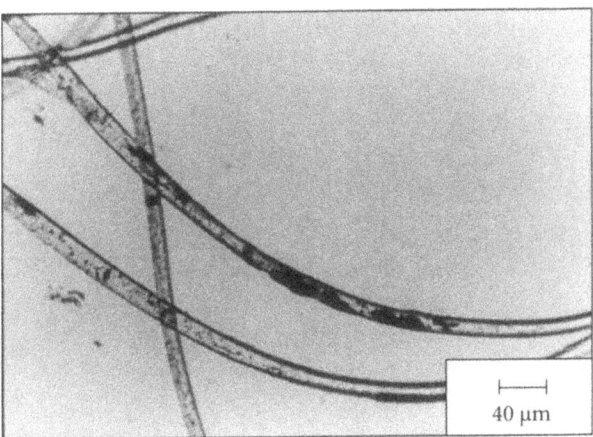

Fig. 267. Polyester fibers from the curtain fabric in Fig. 266. There are rubber deposits which were destroyed during the chlorite bleach.

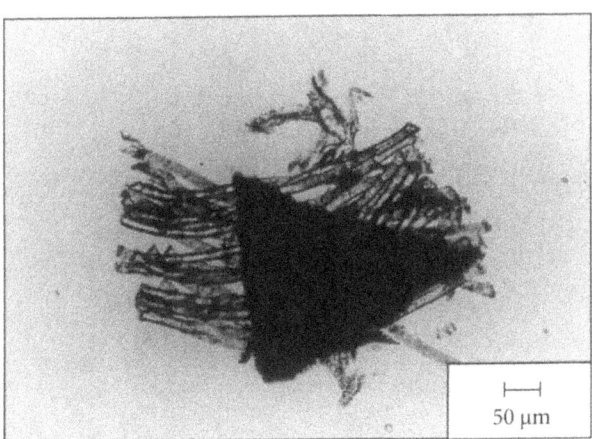

Fig. 268. Polyester fibers from the curtain fabric in Fig. 266, bonded by a film of black marking ink.

ink is resistant to alkaline and acid scouring. It also proved to be resistant to chlorite bleaching. Thus it could not be the cause for the specks. For final verification, the areas with the black marking ink were removed from one bleach lot, but the black specks still occurred.

A check in the plant revealed that the damage had been caused by the fastening wheels of the dyeing beam which were made of black rubber. They had been damaged during chlorite bleaching. The rubber had slowly detached and had been "filtered off" by the fabric. In the subsequent drying process the rubber residues formed a film-like bond with the polyester fiber.

8 Other Defects in the Quality of Textiles

Chemical, mechanical and thermal damage, described in the preceeding chapters, and biological damage that will be dealt with in chapter 9, has recurring causes which (as experience shows) can almost certainly be detected by microscopic examination. In practice, other defects occur which can be caused by various chance events. To determine the causes of these defects in finished textiles, considerable experience, a "criminalistic" intuition and thorough examination of the company-internal conditions are necessary. In these cases, the sense of achievement of the textile microscopist is especially significant. Examples will demonstrate some practical problems encountered.

8.1 Skittery Dyed Wool Yarn – Practical Example

Problems arose during dyeing because wool yarn was dyed skittery. In spite of thorough preliminary scouring and careful dyeing, the defect could not be eliminated.

Figure 269 shows the cross-section of a skittery, red-dyed yarn. Some fibers are properly penetrated while on other fibers only the outer areas of the cross-section are dyed. Obviously, wool qualities of different dye substantivity, i.e. of different origin, had been processed.

Fig. 269
Page 180

8.2 Uneven Wool Printing – Practical Example

A printed wool fabric lacked the desired levelness. The fabric cross-section, Fig. 270, reveals that the printing dye unevenly penetrated into the fabric.

Fig. 270
Page 180

Another lot was printed by adding a wetting agent, which had a favourable influence both on levelness and penetration. This is clearly shown in the fabric cross-section, Fig. 271.

Fig. 271
Page 180

This case is of special interest since it demonstrates that the effect of wetting agents on fabrics can be graphically illustrated by means of microscopic examination. It is even possible to specify the penetration numerically by measuring the dyed fabric sections.

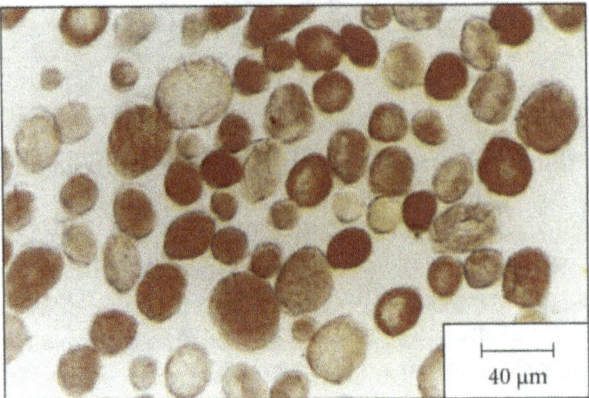

Fig. 269. Cross-section of a skittery dyed wool yarn.

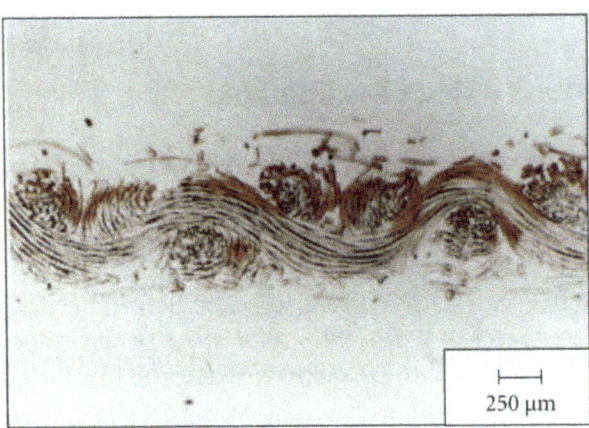

Fig. 270. Cross-section of a printed wool fabric which was printed without adding a wetting agent.

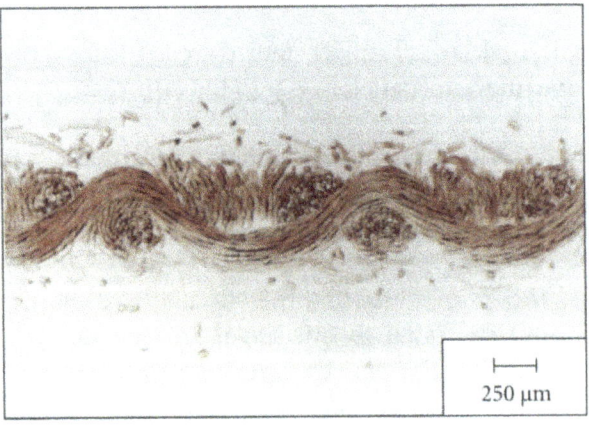

Fig. 271. Cross-section of a printed wool fabric which was printed with a wetting agent.

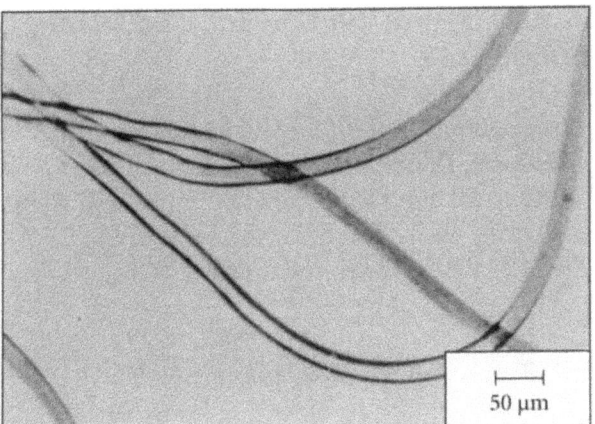

Fig. 272. Polyamide fibers which were reserved as a result of trapped air during the dyeing process.

Fig. 273. Darker dyed areas in a polyester fabric caused by water drops on the fabric before setting.

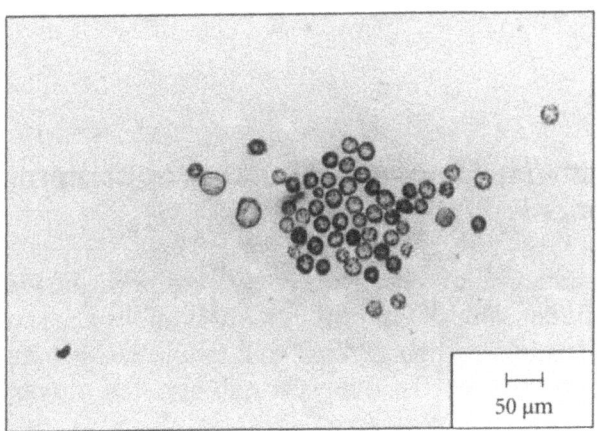

Fig. 274. Fiber cross-section of a yarn made of acrylic/wool. The wool fibers (larger cross-section) lie at the periphery of the thread.

8.3 Small Light Spots Caused by Trapped air Bubbles During the Dyeing of Wound Packages

Fig. 272
Page 181
Similarly to yarn lubricant residues, air bubbles trapped in the material can cause reserving during dyeing. Such textile damage always occurs in wound-up fabrics when it has not been wetted and deaerated properly. Figure 272 shows polyamide fibers which are locally not dyed . The fibers are clean and without any recognizable deposits. The fabric could be repaired by stripping and redyeing.

Suitable wetting agents can avoid such damage from the outset. The heating process should not be performed too fast; it must be ensured that the trapped air can escape before the dye adsorption starts.

8.4 Dye Unlevelness in Polyester Knitwear Caused by Water Drops – Practical Example

Fig. 273
Page 181
Knitwear made of texturized polyester showed dark spots after dyeing, which occurred only in some lots. The dark areas looked like water spots, Fig. 273. Microscopic examination provided no hints of the cause of unlevelness. The fiber material was clean and showed no encrustations or deposits. When looking for the cause of the defect the following observations were purely accidental: the prescoured and dried fabric was stored under a shed roof. Rain drops occasionally fell onto the fabric through an open window. The fabric which was thus too wet in some areas was set on the tenter frame in the usual way. After dyeing – due to changed setting conditions – the rejected stains appeared in the wetted areas. These conditions could be exactly reproduced in a model test.

8.5 Dye Unlevelness Within a Spinning Lot Due to Separation of the Fiber Components – Practical Example

Yarn made of acrylic and wool was dyed on rocket bobbins. One of the spinning lots showed different color shades; on the rocket bobbins the yarn was partially dyed darker, partially dyed lighter from knot to knot. When the yarn was unwound, the lighter yarn layers were yellowish and the darker ones reddish.

Cross-sections were prepared of the lighter and darker yarn layers; microscopic examination revealed that in both yarn layers the wool part was yellowish while the acrylic fiber material was reddish-brown. In the light yarn most wool fibers were found in the outer areas of the yarn, Fig. 274; in the darker dyed yarn, wool fibers were predominantly discovered in the core, Fig. 275. The different dye absorption of wool and acrylic accounted for the shade differences. Since it is impossible to dye wool and acrylic tone-in-tone, dyeing unevenness will always occur in the case of an uneven fiber distribution in the yarn cross-section.

Fig. 274–275
Page 181/184

8.6 Darker Specks on Dyed Feather Bed Ticking Due to Non-Decomposed Seed Husks – Practical Example

Red-dyed cotton feather bed ticking was rejected because of numerous small dark to black specks. It had to be clarified whether textile damage had been caused by seed husks or by impurities such as tar, carbon black or graphite.

Figure 276 illustrates an isolated yarn sample with a speck. Figure 277 shows the same area in stronger magnification. It can be clearly seen that the speck is also dyed. In the speck, the roots of the cotton fiber can be clearly recognized. The specks were caused by seed husks and not, as was presumed, by soilings, such as tar, carbon black or graphite.

Fig. 276–277
Page 184

8.7 Spots Due to Dark-Dyed Fly Fibers – Practical Example

A jersey fabric made of textured polyester yarn showed small dark areas. This was attributed to oil and/or grease soiling; however, the stains could not be removed by dry-cleaning.

Microscopic examination revealed dark-blue and red dyed polyester fibers in the areas in question, Fig. 278, which could only have been caused by fly fibers.

Fig. 278
Page 185

8.8 Red Specks Due to Dyed, Melted and Flat-Rolled Man-Made Fly Fibers on a White Viscose StapleFabric

Specks can have many causes. After heating during drying, singeing or calandering in the finishing process natural fibers or cellulose regenerated fibers can

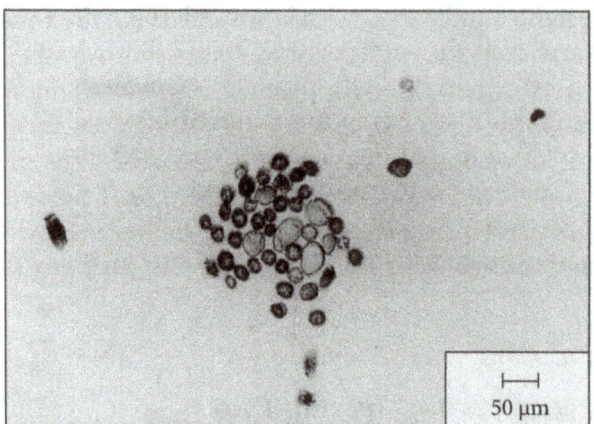

Fig. 275. Fiber cross-section of the yarn shown in Fig. 274. Wool fibers lie in the center of the thread.

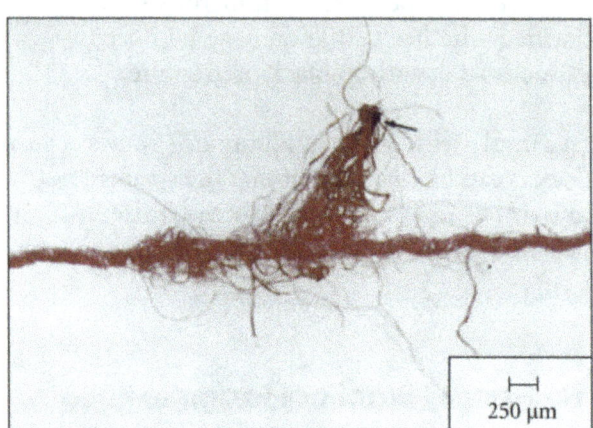

Fig. 276. Cotton yarn with dark speck.

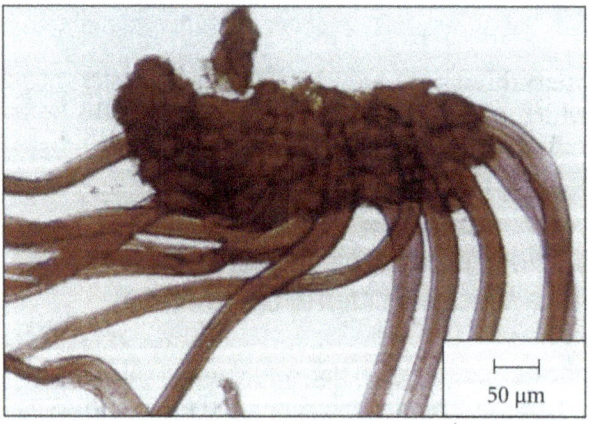

Fig. 277. Strongly magnified speck from Fig. 276. A seed husk with cotton fibers.

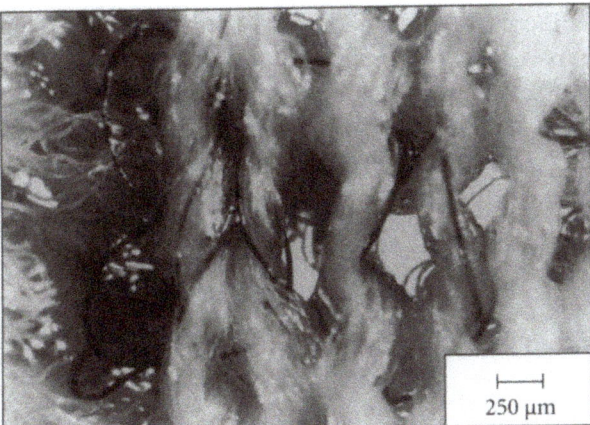

Fig. 278. Spots caused by dark-dyed fly fibers.

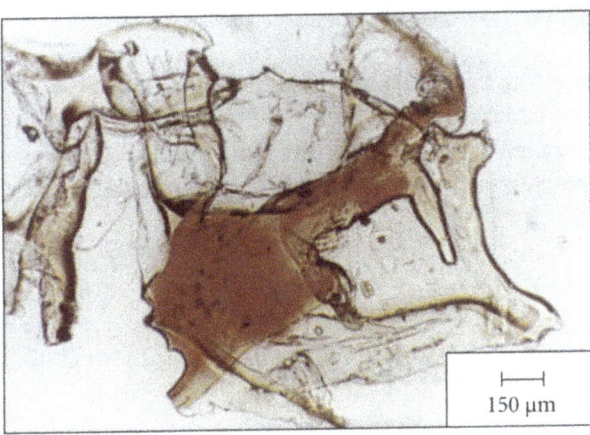

Fig. 279. Colored speck from a white viscose staple fabric; dyed man-made fly fibers which started to melt during finishing due to heat treatment; while running over rollers, the fly fibers were rolled into a film under stress and pressure.

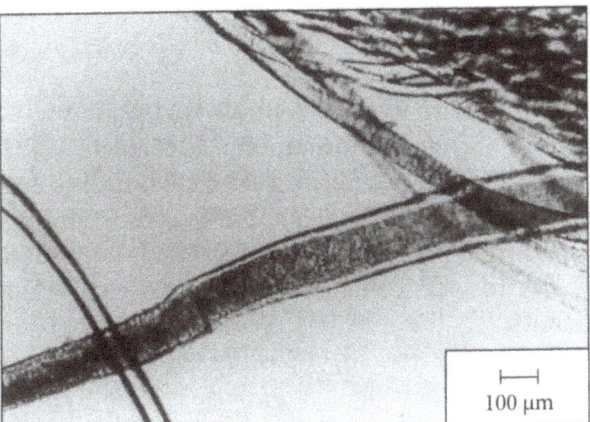

Fig. 280. Short bristly fiber (broad, with medulla) in a wool yarn.

Fig. 279
Page 185
contain melted particles; these particles can additionally be rolled out to form a thin film when running over rollers under stress and pressure. This hints at synthetic fly fibers with a low melting point, such as polypropylene. Figure 279 shows such a (red-dyed) speck isolated from a white viscose staple fabric. It can be clearly perceived that several dyed thermoplastic man-made fibers were melted and squeezed together to form a film.

8.9 Light, Undyed Short Bristly Fibers in a Wool Yarn – Practical Example

Yarns made of pure wool showed light, undyed fibers on the cone surface. Wool had been top dyed on the backwashing machine. The light fiber fractions could allegedly not be recognized on the dyed top.

Fig. 280
Page 185
Microscopic examination of the rejected yarn samples revealed that the fibers which were either slightly dyed or undyed were short bristly fibers. They are strikingly broad with a strong medulla. While the fine wool fibers were thoroughly dyed in the rejected yarn, the short bristly fibers had hardly taken up any dye, Fig. 280.

8.10 Gray, Dot-Like Stains on Needlefelt Sheets – Practical Example

Fig. 281
Page 187
Needlefelt sheets made of polyamide and acrylic showed small gray, dot-like stains. Microscopic examination showed that these were steel gray fibers which, in some areas, had conglomerated in the form of tufts, Fig. 281.

Fig. 282
Page 187
During immersion in zinc chloride-iodine solution, these fibers, unlike polyamide fibers (crenellation reaction: formation of characteristic fiber constrictions after heating the sample, Fig. 282) and acrylic fibers (dissolution), showed no reaction [2]. With a small magnet they were examined for magnetic attraction. This test was positive. These steel fibers were used for the improvement of the electrostatic behavior in the needlefelt. However, they were unevenly distributed in the felt. If steel fibers are sought at the outset, they can be detected more reliably by singeing the surface of a sample with a gas flame. Fine steel fibers light up intensively when smouldering.

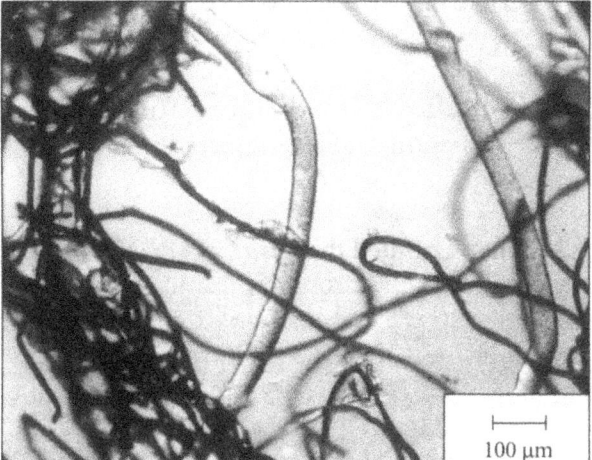

Fig. 281. Polyamide fibers (coarser titer) and metal fibers which have conglomerated into small tufts in a needlefelt.

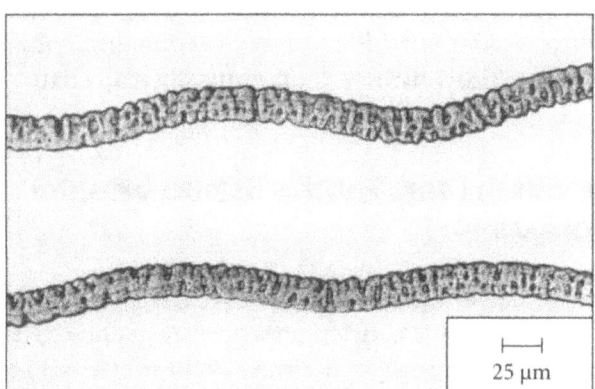

Fig. 282. Crenallation reaction of polyamide fibers: After immersion in zinc chloride-iodine solution and moderate heating, polyamide fibers show typical constrictions similar to terry cloth.

Fig. 283. Film imprint of a cotton tricot fabric with dark streaks. Cause: Stitch displacements in the area of the running marks.

8.11 Running Marks in a Cotton Tricot Fabric
– Practical Example

After alkaline prescouring, mercerized cotton knitwear (tubular fabric) was dyed with direct dye. A streak formation could be observed which was attributed to differences in mercerization.

Fig. 283–284
Page 187/189 However, microscopic examination refuted this assumption. The streaks could also be recognized on the film imprint. Microscopic examination showed that in the outer zones of these darker streaks, threads were tighter and the stitches more open than in the remaining piece, Fig. 283; in the slightly lighter streaks, however, the threads were more open, more voluminous, slightly compressed and the stitches appeared more closed, Fig. 284. Incident light is therefore differently reflected, creating the impression of lighter and darker streaks.

The distorted and/or shifted stitches observed are typical of running marks. Running marks undoubtedly accounted for the microscopic findings. Unlike crease marks (see chapter 3.3.6) the fibers were not mechanically damaged.

8.12 Shade Differences in Dress Fabrics Due to Different Hairiness – Practical Example

A black-dyed dress made of polyamide/wool with white effect threads of polyamide/viscose staple showed clear shade differences on the left and right side of the seam. The right side looked fresher and clearer than the left. The white effect threads were more strongly accentuated on the right fabric part.

Fig. 285
Page 189 For better evaluation, the lengths of fabric to the right and left of the seam were folded and the crease marks were examined under a low-magnification microscope. Figure 285 shows that the fabric surface of one fabric length is hairier than the other one. On the hairy side, the white effect threads were partially covered. As a result, the fabric appeared duller while the effect threads were more distinct in the less hairy fabric part.

8.13 Brittle, Dope-Dyed Acrylic Fiber Material on the Carding Machine and Drawing Frames – Practical Example

A fiber blend was to be examined which consisted of 20% polyamide fibers dyed with 1:2-metal complex dyes, 30% acrylic fibers dyed with basic dyes and

Fig. 284. Film imprint of intact, lighter streaks of cotton knitwear in Fig. 283. Here, the appearance of the goods is homogenous.

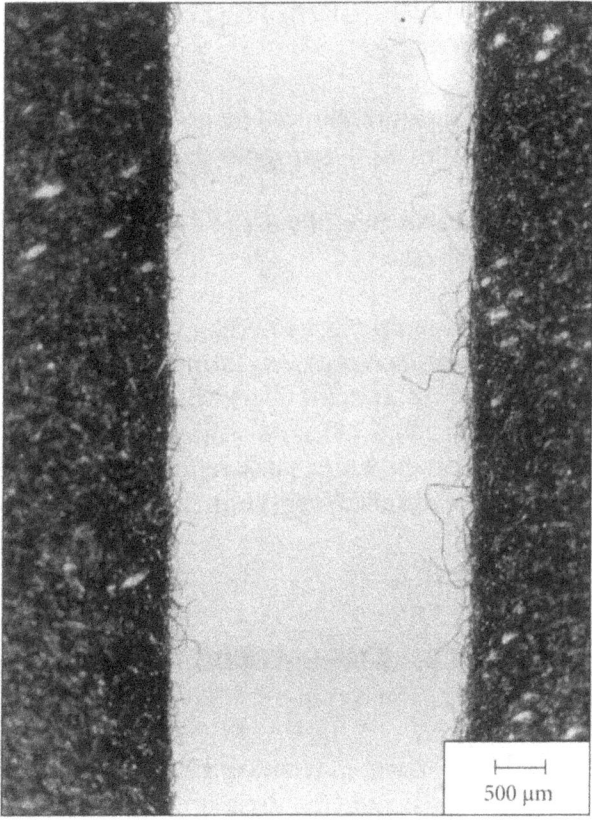

Fig. 285. Dress fabrics with shade differences due to hairiness. Left: Smooth surface, fresh, clear appearance. Right: Hairy surface, dull appearance.

Fig. 286
Page 192
50% dope-dyed acrylic fibers, that had been spun by means of semi-worsted spinning. On the carding machine and the drawing frames, short-fibered, brittle fiber waste resulted, Fig. 286. The acrylic fibers dyed with basic dyes had allegedly been improperly dyed and had been damaged in the high-temperature dyeing unit.

Fig. 287
Page 192
Microscopic examination of the waste revealed that it consisted solely of acrylic fibers. It was soluble in a zinc chloride-iodine solution. Polyamide fibers would have shown the crenellation reaction in this solution, see chapter 8.10, Fig. 282. In order to distinguish between the basic-dyed and the dope-dyed acrylic fibers, concentrated sulphuric acid was dripped onto the fiber samples, Fig. 287. A comparison showed that the dope-dyed acrylic fibers turned yellow. The basic-dyed acrylic fibers did not change the shade, they remained red. The waste taken from drawing frame and carding machine also turned yellow. Consequently, it was proved that material damage was caused by the dope-dyed acrylic fibers and the dyeing with basic dyes had been carried out properly.

8.14 Light Specks in a Milled Terry Towelling Fabric Due to Dead Cotton – Practical Example

Fig. 288
Page 192
After dyeing, a milled terry towelling fabric showed many light or white specks, Fig. 288, whose formation was attributed to improper sizing.

A section of the untreated material was carefully desized and dyed with a reactive dyestuff. It also showed light specks.

Fig. 289–290
Page 193
The specks were also found on a film imprint, Fig. 289. Microscopic examination showed that the rejected specks had not been formed by encrustation or deposits, but that there were broad transparent fibers which had conglomerated to form tufts in the affected areas. Fig. 290 shows a fiber sample with a light speck. The completely transparent fibers with a weak, hardly thickened cell wall and the wrinkles and foldings must unmistakably be attributed to dead cotton, see chapter 2.3.1, Fig. 68.

8.15 Bonded Yarns in a Cotton Cross-Wound Bobbin – Practical Example

A cotton single yarn blended with viscose staple could not be perfectly unwound after pretreatment on cross-wound bobbins. In the areas where the threads crossed on the bobbin they were bonded, Fig. 291. Microscopic examination revealed that in these areas dead cotton fibers had conglomerated to form tufts, cf. Figures 68 and 290. As a result, the yarns occasionally stuck together, thus causing thread breakage during rewinding.

Fig. 291
Page 193

8.16 Knitted Goods Sticking Together in Garment Production Due to Hairiness of Cotton Yarn – Practical Example

A tricot fabric dyed in the yarn caused problems during garment manufacturing since the individual parts stuck together when lying on top of the others. They could only be separated with difficulty. A defective softening treatment was presumed to be the cause. This was examined by extracting a section of the rejected fabric with petroleum ether/ethanol. After this extraction – the softener should then have been removed – the effect reappeared with the same intensity.

According to microscopic examination, the tricot material was made of pure cotton. There were neither incrustations, deposits nor any fiber damage.

The cotton yarn which was isolated from the knitwear showed a relatively marked hairiness. Moreover, the fibers protruding from the yarn bundle were occasionally tangled.

When these yarns were laid on top of each other they could only be separated with difficulty. This was due to the fact that the fibers protruding from the yarn bundle became tangled and twisted, Fig. 292, thus causing an adhesive effect.

Fig. 292
Page 194

8.17 Cotton Fabrics With Side-to-Center Shading Due to Uneven Squeezing Effects – Practical Example

A cotton fabric which had been padded with vat dyes showed a clearly lighter shade at the edges than in the center. At first, incorrect pretreatment was assumed to have caused the different wetting behavior of the edges and the center. This, however, was not true.

Figure 293 and 294 illustrate that the cross-sections taken from the center of the piece are round; cross-sections taken from the edges are clearly flatter and/or more oval. This leads to the conclusion that, when padding with vat dyes, the squeezing pressure on the fabric was clearly higher at the lighter edges than in the center of the piece.

Fig. 293 – 294
Page 194

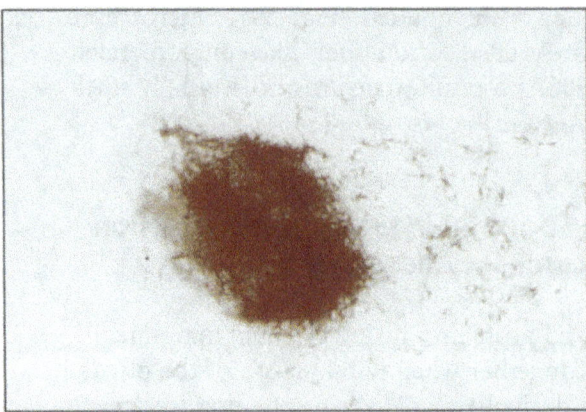

Fig. 286. Waste material from the carding machine and drawing frames during processing of a blend made of polyamide fibers as well as acrylic fibers, which were partially basic-dyed, partially dope-dyed.

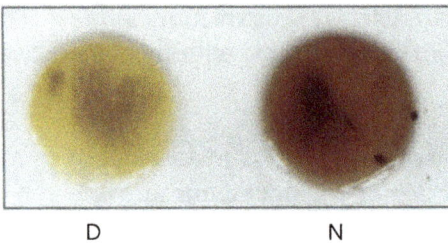

D   N

Fig. 287. Sulphuric acid test on acrylic fibers.
D = Dope-dyed fibers
N = Normal, i.e. basic-dyed fibers,
S = Waste material from the drawing frame,
K = Waste material from the carding machine.

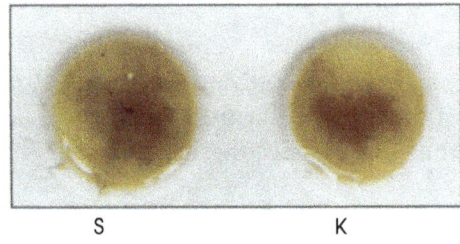

S   K

Fig. 288. Milled terry towelling fabric with light specks after dyeing.

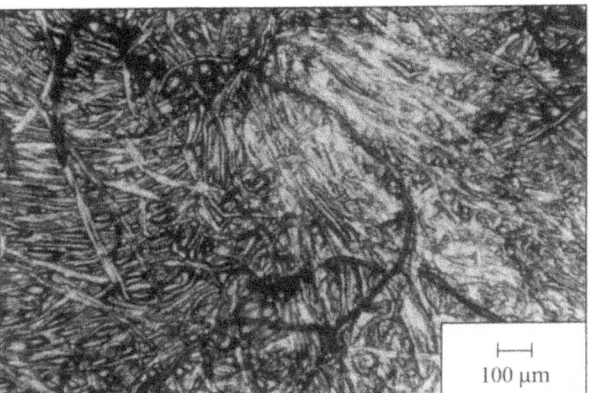

Fig. 289. Microscopic magnification of the film imprint of the milled terry towelling fabric in Fig. 288 with a light speck.

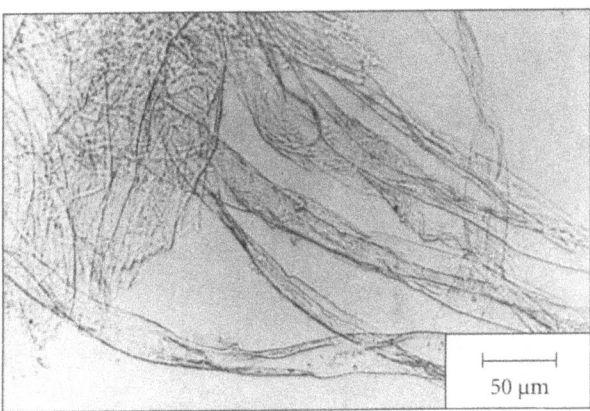

Fig. 290. Fiber sample from a light speck, Fig. 288. Dead cotton can be recognized by its thin cell wall and transparency.

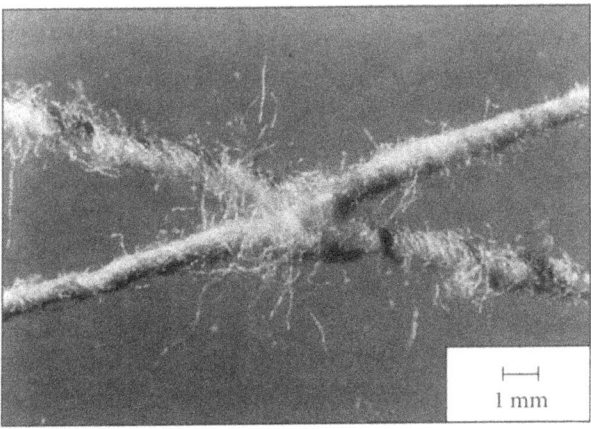

Fig. 291. Yarns sticking together on a cotton cross-wound bobbin. Cause: Neps from immature and dead cotton fibers.

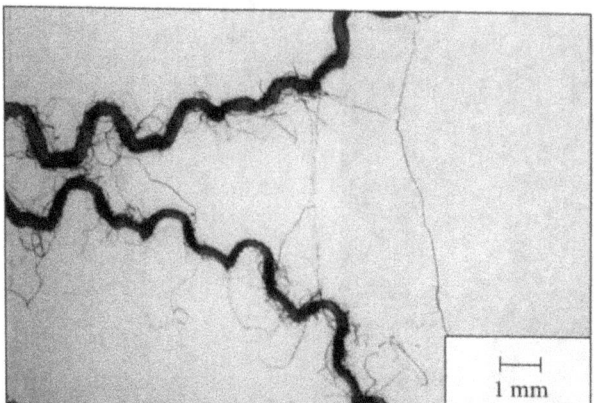

Fig. 292. Hairy cotton yarn, threads sticking together due to tangling and twisting of fibers which protrude from the yarn bundle.

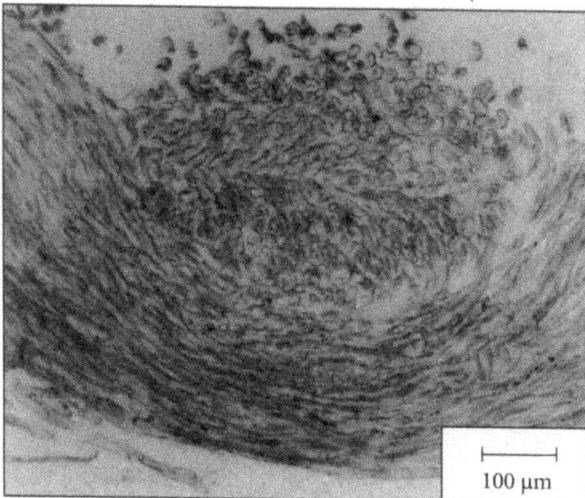

Fig. 293. Side-to-center shading in a cotton fabric. Fabric cross-section taken from the center of a piece. Yarn with almost circular cross-section.

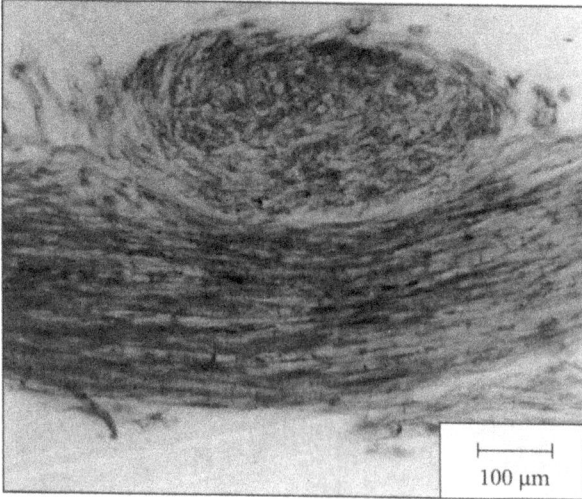

Fig. 294. Side-to-center shading in a cotton fabric. Fabric cross-section taken from an edge of the same piece as in Fig. 293. Yarn deformed into an oval shape due to increased squeezing pressure.

9 Microbiological Damage to Fibers

All natural fibers can be damaged by microbial attack; the fibers themselves serve as a nutrient substrate for microorganisms. In the case of synthetic fibers, the lubricants, sizings, softening agents and finishes used during textile manufacturing assume this function [51]. Favorable preconditions for growth of microorganisms are particularly created by the combination of heat and high atmospheric humidity during storage in still air. During summertime, the hazard of damage by microorganisms is very high in sizing rooms, dyehouses and finishing plants because high room temperatures and high relative humidity prevail in the wet areas. Textiles with an excessive moisture content are frequently packaged in sheets or plastic bags. On the other hand, bottlenecks occuring at drying aggregates in the finishing plants often account for the fact that pieces which are spin-damp or improperly dried must be left over the weekend or the works holidays [52].

9.1 Damage Caused by Fungi

Moulds (Penicillium and Aspergillus types) represent a special hazard [53] to textiles. The spores of these fungi are widespread in the air. If they come into contact with textiles and adhere, they form a dangerous focus of infection. In this context, the pH value of the fabrics is also important. The growth of moulds requires a neutral or, preferably, mildly acid environment.

Damage to textiles caused by moulds can be mostly recognized by the formation of differently colored stains which can be black, yellow, reddish brown, olive-green or orange, depending on the mould type. They are referred to as mildew. A characteristic musty odour is another indication of attack by moulds. However, this odour can be quickly eliminated by exposing the fabric to fresh air. Identification of the stains is then only possible by means of microscopy.

Fungi can be recognized relatively easily under the microscope. However, they are hard to detect if the fabric was washed after the infestation and only residues of fungal attack remain on the fabric. Under these circumstances, residues of fungi can only be detected by means of a staining method. This is especially true if the fungus itself has either been barely dyed or has not been dyed at all, thus being translucent. The Cotton Blue-lactophenol solution according to

Nopitsch [11] has proved highly effective for the dyeing of fungi. Today, Cotton Blue II from the former company Theodor Schuchardt is no longer available, but, according to Bigler [54] both Water Blue B (Hoechst) or Lanaperl Blue RN 150 according to specifications of Hoechst can be utilized. The staining reagent is produced as follows:

solution I   20 ml lactic acid,
                   20 g phenol,
                   40 ml glycerol,
                   20 ml dist. water.

solution II   dissolve
                   2 g dyestuff in
                   100 ml dist. water.

The ready-for-use reagent (so-called Cotton Blue-lactophenol) is made of 50 ml of solution I and 10 ml of solution II.

To detect moulds, the fibers to be examined are placed on a microscope slide and immersed in the dye solution for some minutes. Since the reagent has a dark blue color, blue dyed fungi do not always clearly contrast with the environment. It is therefore useful to suck solution I or dist. water through the preparation with the aid of a filter paper strip (rider) placed at the side of the cover glass. The moulds can now be better recognized.

Another dyeing method for moulds is the application of a mildly acetic acid 0.5 % solution of Methylene Blue B (Merck) that is applied in the same way. Practical examples of the test for fungal attack are described below.

9.1.1 Mould Attack on Cotton – Practical Example

Fig. 295
Page 197
After finishing, a cotton fabric showed olive-green and yellowish stains, Fig. 295. In a blank vat the olive-green tint turned to yellow and regained the original tint after slight oxidation with hydrogen peroxide. This was attributed to stain formation caused by the vat dye utilized in the dyeing process.

Fig. 296
Page 197
However, microscopic examination with Cotton Blue-lactophenol reagent showed that the stains must be attributed to fungal attack, Fig. 296. The stains could be removed by means of chlorine bleach and the fabric recovered totally, since the fungal attack had not caused any loss of strength.

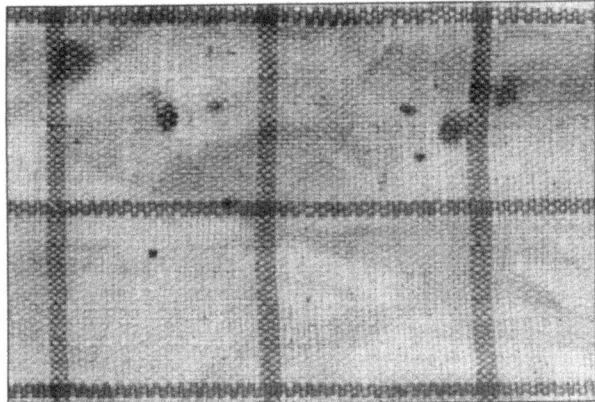

Fig. 295. Moulds on a cotton fabric

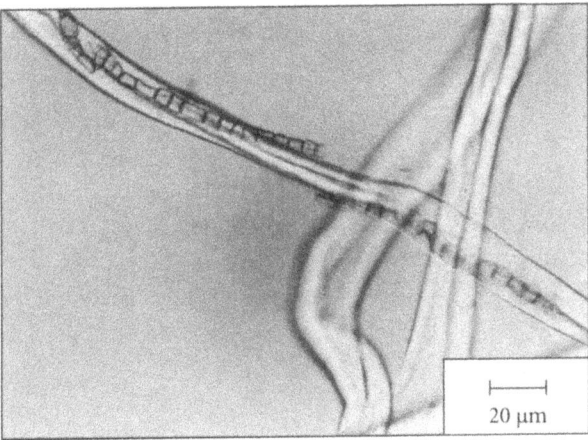

Fig. 296. Moulds on cotton fibers, dyed with Cotton Blue-lactophenol reagent. The fine-structured mycelium threads can be clearly recognized.

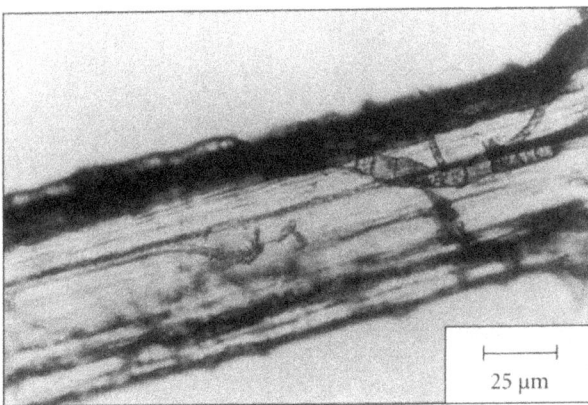

Fig. 297. Moulds on a hemp fiber, dyed with Cotton Blue-lactophenol reagent.

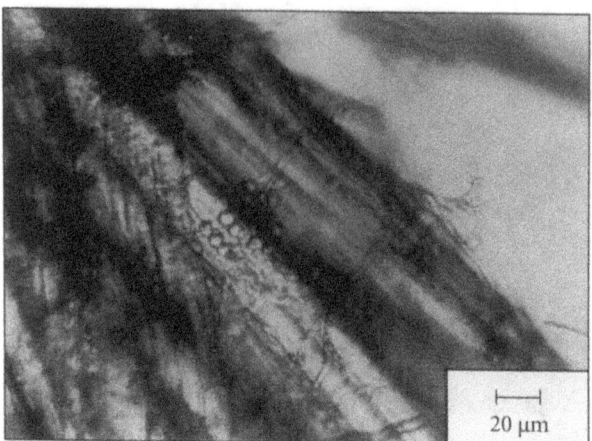

Fig. 298. The same hemp fiber as in Fig. 297 into which mould has deeply penetrated.

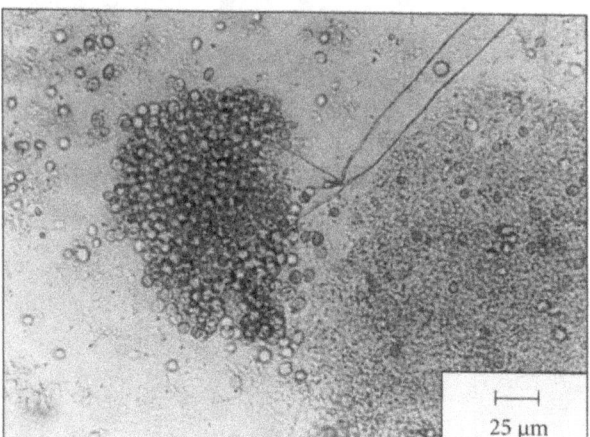

Fig. 299. Stain formation on a hemp yarn caused by moulds.

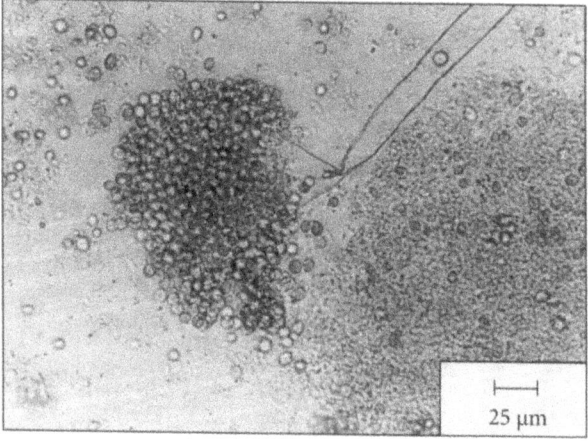

Fig. 300. Mould spread on a hemp fiber.

9.1.2 Mould Attack on Sausage Yarn Made of Hemp
– Practical Example

A sausage yarn sample made of hemp which was rejected due to its low tear strength was to be examined. The yarn showed stains; the tear strength was found to be especially low in these areas.

Microscopic examination revealed that the yarn was infiltrated with moulds. In the stained areas, mould attack had already damaged the fibers.

Figures 297 and 298 show hemp fibers, dyed with Cotton Blue-lactophenol, into which moulds have already deeply penetrated.

Fig. 297 – 298
Page 197/198

9.1.3 Mould Attack on Packing Cords Made of Hemp
– Practical Example

There were several clews of hemp threads on which a strong stain formation of different colours could be detected, Fig. 299.

Fig. 299
Page 198

Microscopic examination revealed that this must be attributed to mould attack, Fig. 300. The fiber material had already been damaged, which was noticeable in a diminution of tear strength in the infested yarn areas.

Fig. 300
Page 198

Experience has shown that mould on hemp fibers almost always leads to a certain reduction in tear strength.

9.1.4 Mould attack on Viscose Staple – Practical Example

Small dark to black stains had formed on a viscose staple fabric. It also showed reddish and brownish stains in some areas. These had presumably occurred during desizing of the fabric.

Microscopic examination showed that the fabric was infested by a dark-colored fungus, Fig. 301.

Fig. 301
Page 200

Further inspection in the plant revealed that the sized warp had not been dried properly; during the works holidays it had been stored for some time in a highly moisturized state in a dark, non-ventilated room in a moist, warm atmosphere. These were ideal conditions for the formation and spreading of fungi.

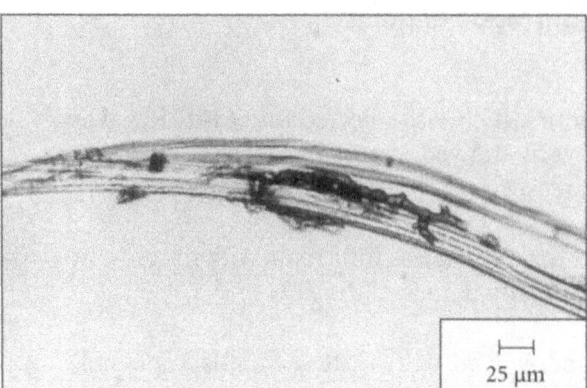

Fig. 301. Viscose staple fiber with dark-colored fungus.

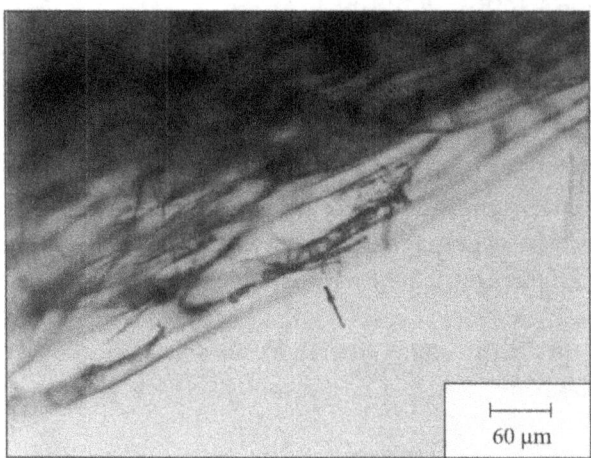

Fig. 302. Residues of a fungus on a polyester/cotton yarn after scouring treatment, dyed with a Cotton Blue-lactophenol reagent.

Fig. 303. Polyester fabric with dark, spot-like streaks in warp direction, caused by a dark-colored fungus.

As the microbial decomposition of cellulose fibres had not yet progressed to a significant degree of fiber damage, the fabric could be restored by means of chlorine bleaching.

9.1.5 Moulds on a Blended Fabric Made of Polyester/Cotton – Practical Example

After desizing, a yarn-dyed fabric which consisted of equal parts of polyester and cotton showed small light gray stains. This was attributed to inadequately scoured sizing agent (specifically a product made of starch and polyvinyl alcohol). However, examination for sizing residues was negative, the fabric was perfectly desized.

In order to find out whether the stain formation had been caused by other deposits or local structural changes of the fiber material, a large imprint was made on a polystyrene film. The stains could also be recognized on this imprint, although only vaguely. Consequently, they could not be due to dyeing unlevelness. They must be attributed to deposits or local structural changes at the surface.

The Pauly-reagent dyed the light spots orange. Residues of a protein size could have possibly been the reason for this since the Pauly reagent also dyes protein sizes. Microscopic examination, however, refuted this presumption. Both dyeing with Cotton Blue-lactophenol and with Methylene Blue, Fig. 302, showed that it was a residue of hyphomycete.

Fig. 302
Page 200

Later, it emerged that the fabric had been padded with an enzyme liquor for desizing and had been stored over the weekend. Storing a wet fabric at mid-summer temperatures enabled the development and spreading of moulds over the piece. Once the cause of stain formation was known, a chlorine treatment was carried out with 0.5 g/l active chlorine. However, this treatment was not successful. Only a considerably stronger chlorine treatment with 2 g/liter active chlorine for 30 minutes at 25 °C could completely remove the mould spots on the fabric.

9.1.6 Moulds on a Sized Polyester Fabric – Practical Example

A gray-state polyester fabric which had been sized with polyvinyl alcohol was covered with dark streaks, Fig. 303. Preliminary examination revealed that

Fig. 303 – 304
Page 200/202

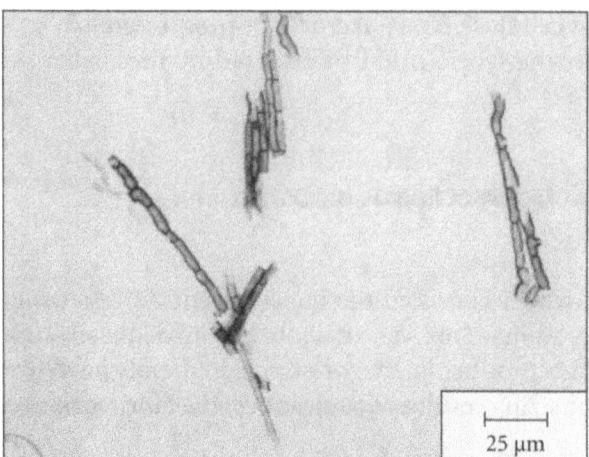

Fig. 304. Dark-colored fungus from Fig. 303, washed off with dist. water from a sized polyester fiber.

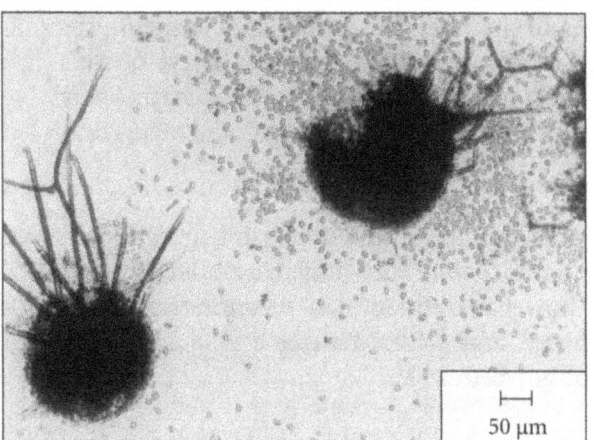

Fig. 305. Mould on the paper label of a polyester yarn.

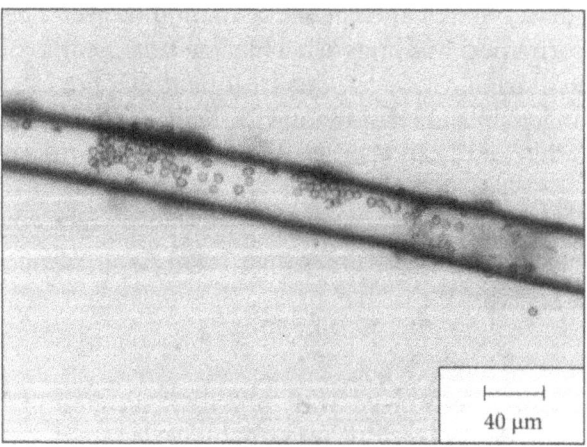

Fig. 306. Fungus spores on a polyester fiber.

the spot formation was limited to the warp threads. Microscopic examination showed an attack by a dark-colored fungus, Fig. 304.

In this case it also turned out that the warp had not been properly dried but had been packaged in plastic sheets in a highly moisturized state, thus supporting the formation of fungi on the polyvinyl alcohol sizing agent. Again, restoration was possible by chlorine treatment of the pieces.

9.1.7 Moulds on a Polyester Yarn – Practical Example

After a long storage time in airtight polyethylene bags, gray-green spots appeared on the paper labels of a polyester yarn. Parts of the labels were already completely destroyed. Near the labels, which felt damp, gray and green stains could be recognized on the polyester yarn.

Microscopic examination revealed mould attack, Fig. 305 and 306. Glue on the labels had apparently served as a nutrient substrate for the microorganisms. Since the resistant polyester yarn was not damaged, a short treatment with chlorine bleaching liquor (1 g/liter active chlorine) could eliminate the damage.

Fig. 305–306
Page 202

9.1.8 Dark Stains on a Wool Fabric Caused by Moulds – Practical Example

On a green dyed fabric made of pure wool, dark stains were observed after long-term storage. Microscopic examination revealed an attack with dark dyed fungi, Fig. 307, on the warp yarn of the fabric. Since the yarn-dyed warp had been sized with a starch derivative, this apparently served as a nutrient source for the fungi. The sized warp material had not been dried properly and had been stored in a moist-warm atmosphere for a long time.

Fig. 307
Page 204

With a warp dried and stored under normal conditions, fungi cannot spread on a fabric sized with a starch derivative. Experience shows that no fungal attack should be expected up to a relative humidity of 50%. With a relative humidity of 75% a rapid spreading of fungi should be expected at summer temperatures.

In wool pieces, the removal of mildew stains creates difficulties. Due to the damage to wool caused by chlorine treatment this need not be considered further. In most cases the fabric can only be redyed black. However, a faultless fabric cannot always be achieved [55].

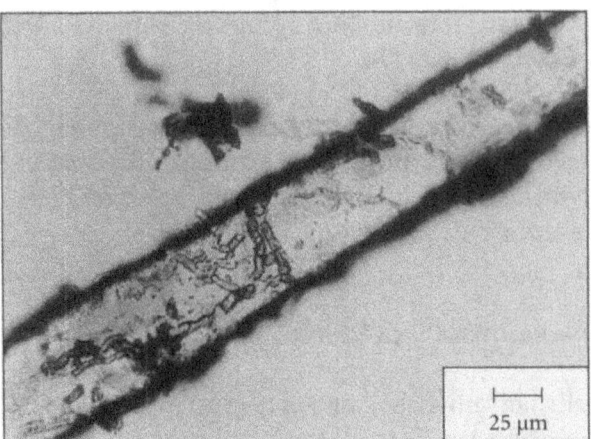

Fig. 307. Wool fiber with mildew, dyed with Cotton Blue-lactophenol reagent.

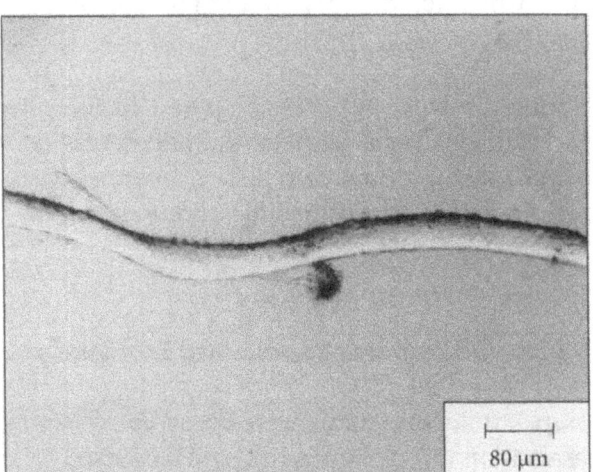

Fig. 308. Wool fiber with mildew; fiber preparation from a dark green stain of a wool yarn.

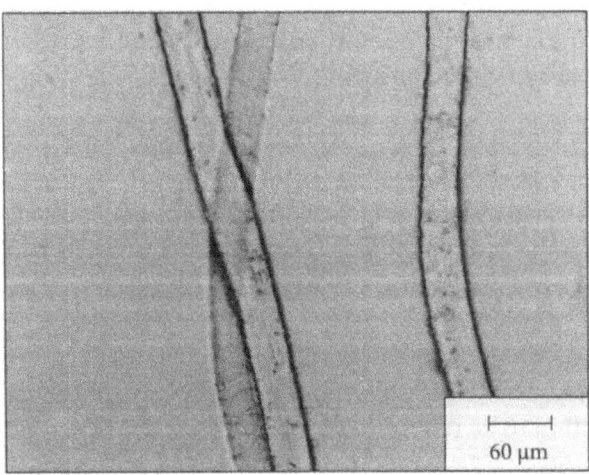

Fig. 309. Wool fiber with mould spores.

9.1.9 Cross-Wound Bobbins Made of Wool with Brownish and Dark Green Mould Stains – Practical Example

After twisting pure wool yarn was steamed, rewound and paraffinized. Post-moistening for weight regulation was subsequently performed. After long-term storage of the cross-wound bobbins, brownish and dark green stains appeared on the wool yarn.

The stained sections in the yarn were examined under the microscope. The examination revealed that all stains were caused by fungal attack, Fig. 308 and 309.

Fig. 308 – 309
Page 204

In order to check the degree of damage to the fiber substance, the Pauly reaction was carried out (see chapter 2.1.1) where wool damaged by moulds is dyed orange to reddish brown depending on the degree of damage. In this case, stained areas were partially dyed red, i.e. wool was badly damaged locally by the fungal attack, which could also be recognized from the partially destroyed scale, Fig. 310.

Fig. 310
Page 206

9.1.10 Warp Beams Made of Polyamide/Wool with Differently Colored Mildew Spots – Practical Example

In a warp made of polyamide/wool 30/70, which had been sized with a starch derivative, lemon, dark brown and reddish brown stains appeared especially in the middle of some warp beams after several days. The fabric also had a strongly musty odour.

Microscopic examination showed that both wool fibers and polyamide fibers were affected by moulds, Fig. 311 and 312. A further check showed that the warp material had not been dried properly. Determination of the moisture content of the warp from different warp beams brought the following results:

Fig. 311 – 312
Page 206

	% Residual moisture
Warp beam, not stained	13
Warp beam, stained	
a) middle	> 30
b) left edge	25
c) right edge	27

It is assumed that the moisture content of the fabric was even higher at the beginning than after storage for several days. Fungal attack was caused by

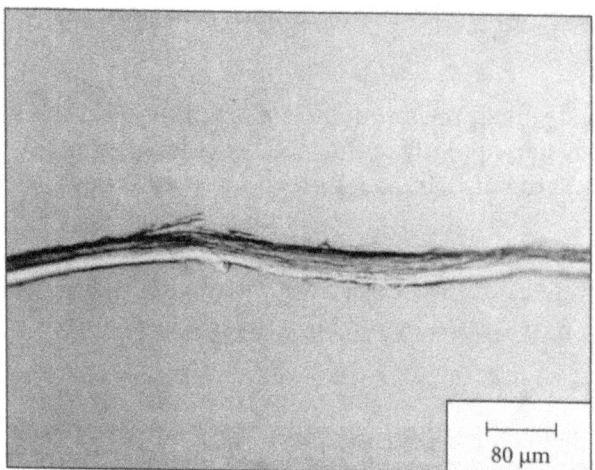

Fig. 310. Wool fiber, damaged by fungal attack.

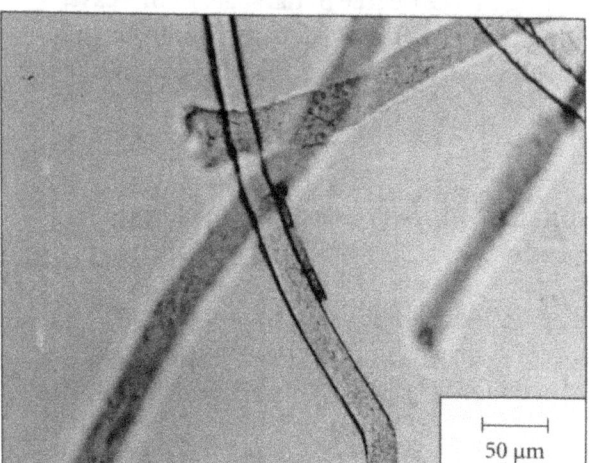

Fig. 311. Dark-colored fungus on a polyamide fiber.

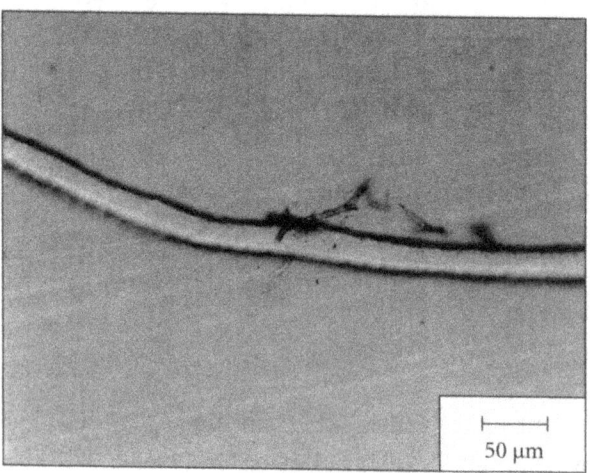

Fig. 312. Dark-colored fungus on a wool fiber.

Fig. 313. Stains in a wool fabric caused by bacteria. Above: Initial sample, untreated.
Below: After treatment with the Pauly reagent. Relatively strong dye absorption shows that the damage to wool by bacteria is already advanced.

storing the warp in a highly moisturized state, especially at summer temperatures for a long period.

9.2 Damage Caused by Bacteria

Animal fibers made of proteins are particularly sensitive to bacterial attack.
Wool-damaging bacteria are predominantly earth bacteria, which can be easily
brought into production plants, for example via dirty shoes [56]. High
moisture content in the fabric, high storage temperatures and a neutral to
mildly alkaline pH value of the fibers are optimal preconditions for the prop-
agation of these bacteria.

Bacterial damage of wool can only be prevented if the spin-moist pieces are
dried as soon as possible. A few practical examples will show that – similar to
fungal attack – considerable damage from bacterial attack can occur when
moist wool is stored over a weekend. If immediate drying is not possible, the
pieces should be well acidified because wool-damaging bacteria do not repro-
duce at low pH values [53, 56, 57]. It was found that wool with an undamaged
scale layer is hardly attacked by bacteria. However, if the scale structure is loose
or pre-damaged, bacteria can very easily penetrate into the spindle cell layer
and degrade the cement substances between the spindle cells. Such pre-damage
can already exist on living sheep, caused by exposure to light, alkaline dust,
excrement or mechanical injuries (mainly of belly wool, see Fig. 31).

Fig. 313
Page 207 Similar to mould attack, bacterial attack can be recognized from different kinds
of stains, Fig. 313.

Fig. 314
Page 210 In gray-state wool fabrics, bacterial attack often cannot be recognized exter-
nally, since numerous wool-damaging bacteria either have no or only little
inherent color [53,56]. Damage can be recognized – mostly only after dyeing
and finishing – by light areas and/or stains, if the wool fibers have been inter-
nally damaged by bacteria and fibrillar dissolution started to appear, Fig. 314.
In spindle cells protruding and/or detached from the fibers, incident light is
refracted, giving the impression of lighter dye absorption or deposits. The
darker the later coloration of the fabric, the more strongly these defects are
visible. In our experience, the fabric will not be examined before the dyeing
process. Through scouring and subsequent treatment with boiling dyeing
liquor, the bacteria colonies which could have been dyed with Cotton Blue-
lactophenol [54] are removed from the fabric, thus rendering detection in the
dyed fabric impossible. Closer examination of the fiber image of bacterially

damaged wool is helpful because the features are so characteristic that an ex-perienced textile expert will have no difficulties in distinguishing it from other damage. Microscopic examination is easier if the damaged areas are first identified by means of dyeing with Pauly reagent or with Cotton Blue-lactophenol.

One typical feature of bacterial attack on wool fiber is an increased longitudinal stripiness of the fibers which indicates the beginning of delamination into spindle cells. In the next step spindle cells are laid open. It is characteristic of bacterially damaged wool that only one half of the wool fibers is decomposed, while the other half seems to be completely intact, Fig. 315 and 316. This must be attributed to the bilateral structure of wool [58]. The ortho cortex cells and para cortex cells differ in their resistance; the ortho cortex cells represent the more labile part of the wool fibers and seem to be more easily attacked by bacteria. The more reactive ortho cortex cells can also be more easily dyed, e.g. with 2% o.w.f. Methylene Blue D in neutral, boiling liquor for half an hour. In the fiber cross-section, they can thus be easily distinguished from the para cortex cells, Fig. 317.

Fig. 315–317
Page 210/211

Acid-damaged wool also shows fibrillar dissolution phenomena. If it is unclear whether there is acid or bacterial damage, this can be resolved by means of a swelling reaction with ammoniacal potassium hydroxide which allows the detection of acid damage to wool even in the initial stage, see chapter 2.1.3, Fig. 47–49. On the other hand, bacterially damaged wool reacts like un-damaged wool in the KV reaction.

Dyeing of wool influences its sensitivity to microbial attack. Wool pre-treated with chromium and/or dyed with chromium dyes is more or less resistant to bacteria [53, 52]. Wool damaged by bacteria, which leaves light spots on dyed fabric, cannot be repaired. Even redyeing is hardly useful in such cases.

9.2.1 Bacterial Attack on a Military Cloth – Practical Example

A green and reddish brown mottled military cloth showed light areas. Stain formation was not yet recognized after the fulling process, where the wet fabric had been rolled up and left over a weekend, and a substantial increase in temperature of the pieces had been found inside the cloth batch. Only after scouring, impregnating and final treatment of the pieces could stains be observed; only green-colored wool fibers had been lightened in the blend.

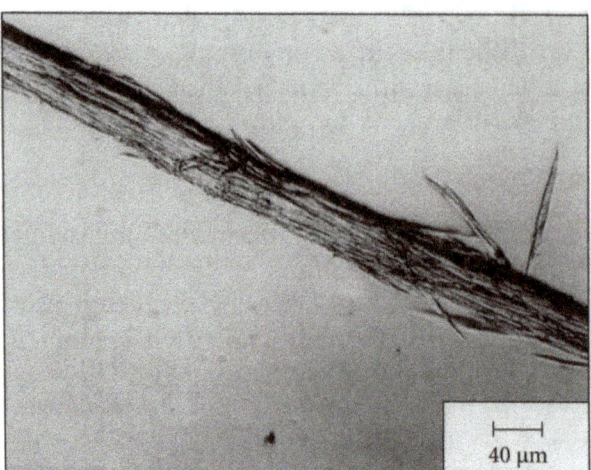

Fig. 314. Wool fiber badly damaged by bacteria with spindle cells laid open.

Fig. 315. Characteristic example of bacterially damaged wool. Often only one fiber half is decomposed while the second half is still undamaged. Such wool is also dyed orange to reddish brown by the Pauly reagent.

 Intact wool

 Step 1: longitudinal stripes

 Step 2: one half decomposed

Step 3: complete decomposition

Fig. 316. Schematical representation of stepwise decomposition of wool caused by bacteria.

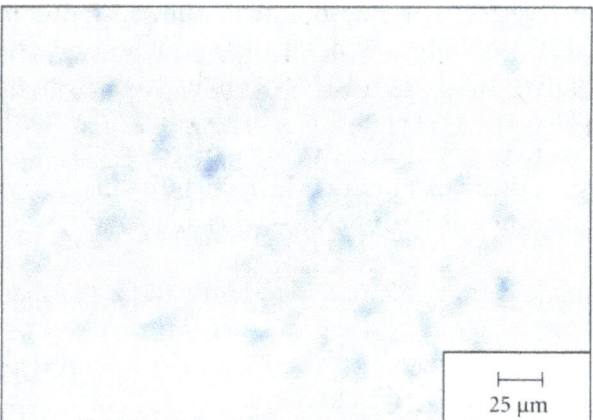

Fig. 317. Detection of the bilateral structure of wool due to dye absorption with Methylene Blue. Ortho and para cortex cells (blue and undyed respectively) can be easily distinguished in the fiber cross-section.

25 µm

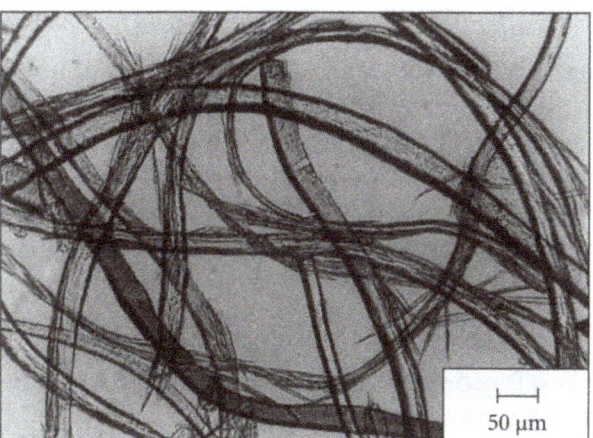

Fig. 318. Bacterially damaged wool fibers from a mottled cloth. The wool fibers dyed with chromium dyes – brown in this case – are not destroyed by bacteria.

50 µm

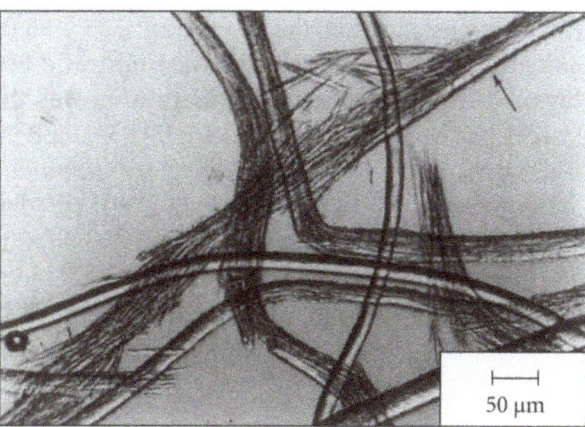

Fig. 319. Wool fibers damaged by bacteria. One half of the fibers attacked longitudinally is partially undamaged (arrow), see Fig. 315.

50 µm

Fig. 318
Page 211
Microscopic examination revealed that the wool fibers showed splintering in the affected areas. Individual wool fibers were virtually dissolved into spindle cells, Fig. 318. Optically lighter stains were produced on these laid open spindle cells. Since the KV reaction (see chapter 2.1.3, swelling reaction with ammoniacal potassium hydroxide) was negative, only bacterial damage could be the reason, especially since often only one half of individual fibers was decomposed longitudinally, whereas the other half was still intact.

Another oberservation made during microscopic examination was that only green dyed wool fibers were bacterially damaged. Later it turned out that the non-affected reddish-brown fibers had been dyed with chromium dyes and therefore showed a higher resistance against bacteria.

9.2.2 Bacterially Damaged Carpet Yarn – Practical Example

Fig. 319
Page 211
In some areas, carpet yarn made of wool with a low proportion of polyamide fibers was stained yellow, pink, olive-green and brown. The rejected areas were over-dyed orange to reddish brown by the Pauly reagent. These areas of carpet yarn were badly damaged. Microscopic examination revealed that the wool fibers in these areas had been dissolved into the spindle cells. Such fiber destruction, Fig. 319, with half the wool in the longitudinal direction of the fibers still intact, is a typical demonstration of bacterial fiber damage.

9.2.3 Streak Formation in a Dyed Fabric Made of Polyester/Wool Caused by Bacterial Attack – Practical Example

Fig. 320 – 322
Page 213
In a fabric section made of polyester/wool, which was dyed dark blue, there were light streaks and stains parallel to the warp, Fig. 320. In some areas the gray-state material showed different stains in the warp direction which were difficult to detect. The stained areas of the untreated material could be dyed reddish brown by the Pauly reagent. Microscopic examination showed that the wool fibers were damaged very badly in the affected areas, Fig. 321 and 322. In addition to wool fibers largely dissolved into spindle cells, there were still intact wool and polyester fibers. In numerous wool fibers, only one half was destroyed in the longitudinal direction, see Fig. 319, while the other half was still intact. Consequently, wool had undoubtedly been damaged by bacteria. The reason was over-moistening of the warp caused by excess application of a warp wax emulsion.

Fig. 320. Fabric made of polyester/wool with light streaks in the warp direction. The streaks must be attributed to bacterially damaged wool fibers, which become visible only after dyeing.

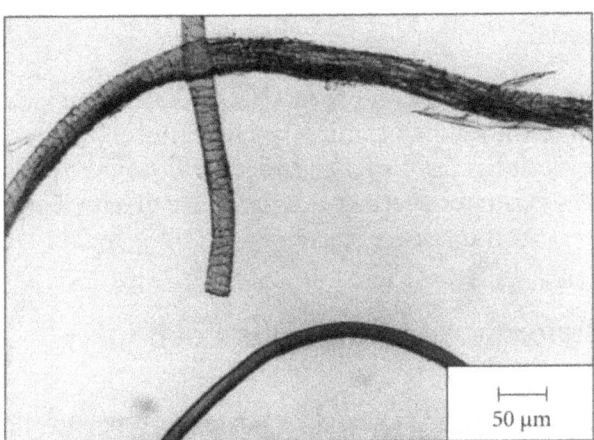

Fig. 321. Fiber preparation of a light streak from the fabric in Fig. 320. Bacterially damaged wool fiber, undamaged wool fiber, intact polyester fiber (below).

50 μm

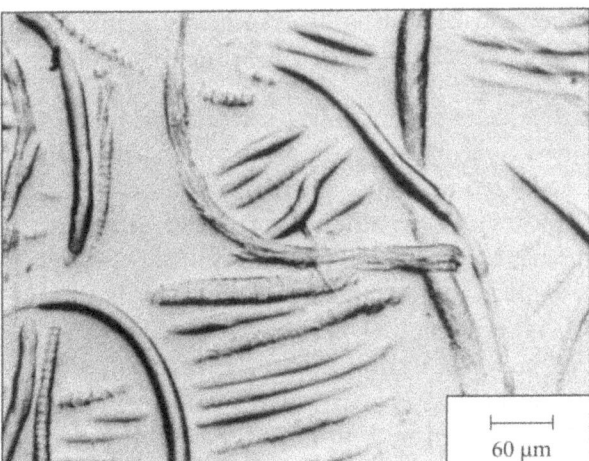

Fig. 322. Film imprint of a light streak taken from the fabric in Fig. 320, in which the partially damaged wool fibers and smooth polyester fibers can be recognized.

60 μm

10 Poultry Feathers as Filling Material for Bedding and Textiles – Analysis of Faults

Poultry feathers give very good protection against the cold. At the present time, they are not only used as filling material for bedding but are also highly sought after as padding material for sleeping bags and garments such as sportswear for mountain climbers, anoraks, padded jackets and coats. These uses also raise the question of care treatments for this material. For example, pillows filled with poultry feathers are subjected to a so-called disinfectant laundering at 95°C when used in hospitals. In this case special finishes and, above all, suitable detergents and laundering processes are necessary in order to avoid drastic damage to the natural product.

Inappropriate laundry treatments and the use of unsuitable detergents can cause marked destruction of the feathers, which, in ignorance of the chemical facts is often attributed to poor finishing of the feathers or down or to the poor quality of the material. It has therefore been found necessary to have simple test methods in order to determine the cause of the damage.

10.1 Chemical and morphological structure of poultry feathers

Poultry feathers consist of proteins, as does wool, and they are thus similar to wool in their chemical properties [59, 60]. Particular mention should be made here of their sensitivity to alkali and chlorine. The structure of poultry feathers is so characteristic that they can be distinguished without difficulty from all other natural products and synthetic fibres used as filling materials.

Figs. 323–325 page 217 All poultry feathers have the same principle of construction. From the round hollow quill the fine, thread-like barbs extend on both sides. These barbs branch out further into so-called barbules. The barbules have hooklets, Figs. 323, 324, 325, which interlock with each other thus forming a network of barbs and barbules which is described as the vane. On account of this principle of construction, poultry feathers contain a large number of air channels between the barbs and barbules. It is this insulating air layer between body temperature and external temperature which results in the even warmth under feather beds and the pleasant feeling of comfort when wearing garments padded with feathers in cold weather. In addition, their hydrophilic properties together with the fine air channels allow good transport of moisture.

Goose, duck, and chicken feathers differ from each other as follows: a squat form is typical for goose feathers, Fig. 326. The lower part of their quill is covered to a greater or lesser degree with a fluffy growth. The quill shows a marked degree of curvature and is as elastic as a steel spring, Fig. 327.

Fig. 326–327
page 218

Duck feathers do not appear as squat as goose feathers and they are somewhat slimmer and more gracefully shaped, Fig. 328. The upper end is tapered. This allows duck feathers to be distinguished easily from goose feathers.

Fig. 328
page 218

Chicken feathers show a number of essential differences compared to goose and duck feathers. Their quill is thinner and much more sensitive to stress. What is particularly noticeable, however, is that their quills are straight and only slightly curved at the top. Chicken feathers therefore do not have the elasticity of goose and duck feathers and do not form as many pores. The feathers practically lie more or less flat beside or above each other. Their filling capacity is therefore relatively low compared to goose and duck feathers. Since chicken feathers enclose less air their insulating properties are also inferior to goose and duck feathers. Quite typical for chicken feathers is the small supplementary feather at the end of the quill, Fig. 329. However, this breaks off easily and is therefore often no longer to be found.

Fig. 329
page 219

In contrast to feathers down does not have a long quill but only a nucleus. Highly branched, silky-soft hairs grow out of this, Fig. 330. Down is very light and has the maximum capacity for storing air. Down is only supplied by geese and ducks and represents the most valuable filling material provided to us by nature.

Fig. 330
page 219

10.2 Detection of damage to poultry feathers

Mechanical damage can be recognized under the microscope, for example in the form of broken or cracked quills. Missing barbs, barbules and hooklets which have been broken off by high mechanical stress are also typical signs for mechanical damage, Fig. 331. Slight chemical damage, e.g. such as that found with feathers that have been somewhat more heavily bleached, cannot be detected by microscopical investigations. Detection can be achieved with the aid of staining tests, which demonstrate all kinds of damage, even in cases where the damage is only slight.

Fig. 331
page 219

10.2.1 Detection of damage to poultry feathers with the Pauly reagent

Fig. 332–334
page 220
On account of their close chemical relationship it seemed obvious to apply staining tests for wool to poultry feathers. It was found that damage to feathers can be detected very reliably with the Pauly reagent. Chemically damaged feathers are stained by the Pauly reagent according to their degree of damage in shades ranging from yellow to orange and reddish brown, as with wool, cf. Section 2.1.1, Fig. 29. Undamaged material remains unstained, Fig. 332. The skin flakes found at the quill end of every feather can also be detected with this reagent, Fig. 333 and 334.

10.2.2 Detection of damage to poultry feathers with Neocarmin W

Fig. 335
page 221
The detection of damage to feathers with the dyestuff reagent Neocarmin W is very simple. The feather sample being investigated is immersed for 5 minutes at room temperature in the dye solution and occasionally briefly stirred. The sample is then rinsed with cold water until the rinsing water is clear. The feathers or down are then centrifuged and air-dried. According to their degree of damage they are then stained as follows, Fig. 335:

undamaged feathers and down:	yellow
skin flakes at the end of the quill:	dark reddish-brown
slight damage:	orange
medium damage:	dark orange to red
heavy damage:	reddish-brown to violet
rotted feathers (biological damage):	dark reddish-brown, partially dirty olive green

The staining test with Neocarmin W has proved to be very useful in practice and because it is easy to carry out it is preferred to the Pauly reagent for analysing damage to feathers.

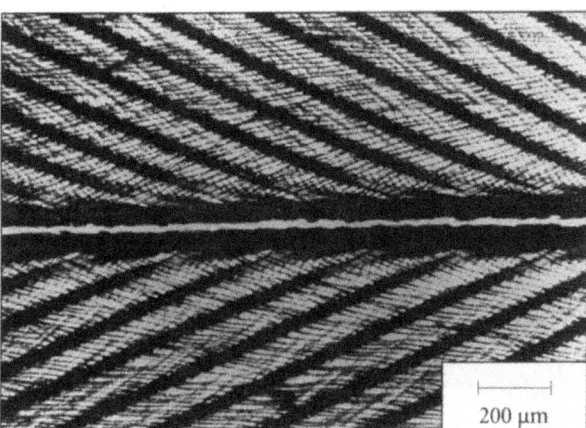

Fig. 323. Section of a poultry feather with the quill in the center and the barbs and barbules extending sideways.

Fig. 324. Isolated barbules from a feather. The hooklets which cause the interlocking of the barbs and barbules can only be seen at higher magnification.

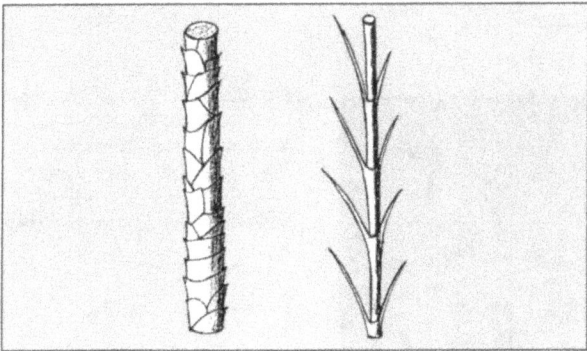

Fig. 325. Schematic diagram of the morphological construction of a poultry feather in comparison to the chemically related wool fiber. In the wool fiber the scales lie close to the fiber surface whereas the construction of the poultry feather resembles a grass stalk. The hooklets, which broadly speaking correspond to the scales of the wool fiber, are of different lengths on each side of the barbule.

Fig. 326. Goose feather with its typical squat form. It looks as if it were cut off at the top.

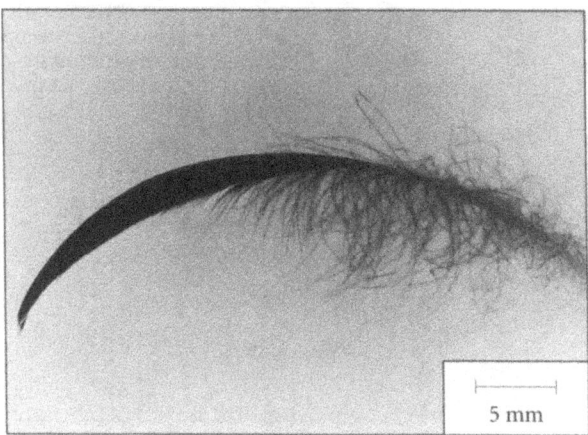

Fig. 327. Goose feather as in Fig. 326, seen from the side. The strong quill of the goose feather has a marked curvature, which results in a high degree of elasticity.

Fig. 328. Duck feathers have a more graceful form than the goose feather. They are tapered at the end. They can thus be easily distinguished from goose feathers.

Fig. 329. Chicken feather with its thin, almost straight quill, which is more sensitive to stress. It is only slightly curved at the top. Also typical is the small supplementary feather at the lower end of the quill. This breaks off easily and is therefore often no longer to be found.

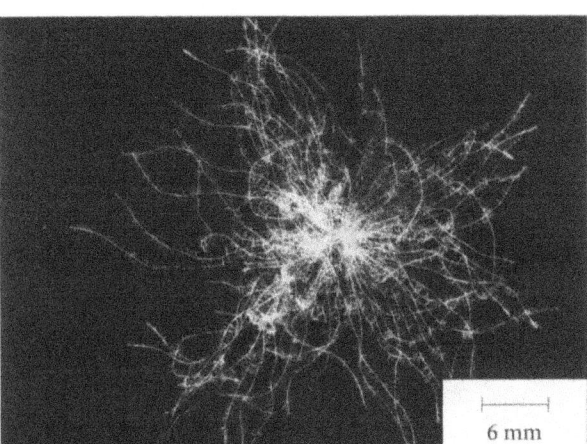

Fig. 330. Only geese and ducks supply down. Down does not have a quill but only a nucleus from which silky-soft hairs grow in all directions.

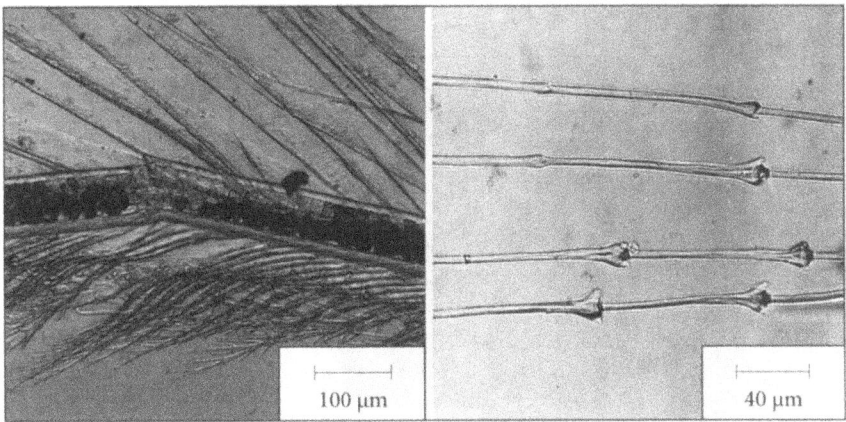

Fig. 331. Mechanically damaged goose feather with a cracked quill. Due to mechanical effects some of the hooklets have also been broken off (photo on the right).

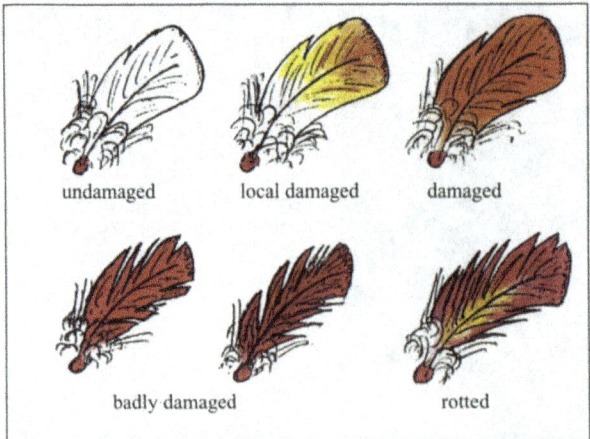

Fig. 332. Schematic drawing of feathers treated with the Pauly reagent. According to their degree of damage they are stained in shades ranging from yellow to orange and reddish brown. (undamaged; local damaged; damaged; badly damaged; rotted).

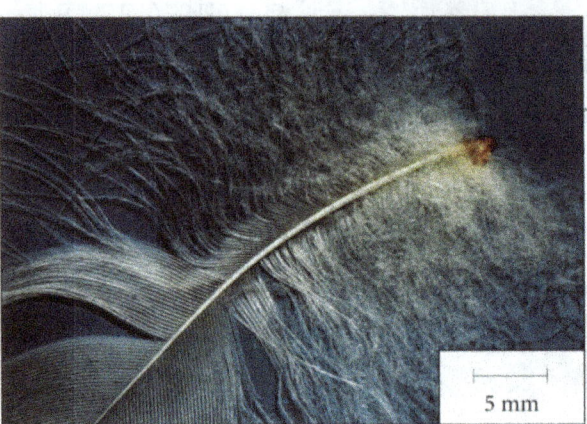

Fig. 333. Goose feather treated with the Pauly reagent. The lower end of the quill is always stained intensive reddish-brown.

Fig. 334. Down treated with the Pauly reagent. On account of its reddish-brown staining, with appropriate magnification the nucleus can be more easily recognized under the microscope than in Fig. 330.

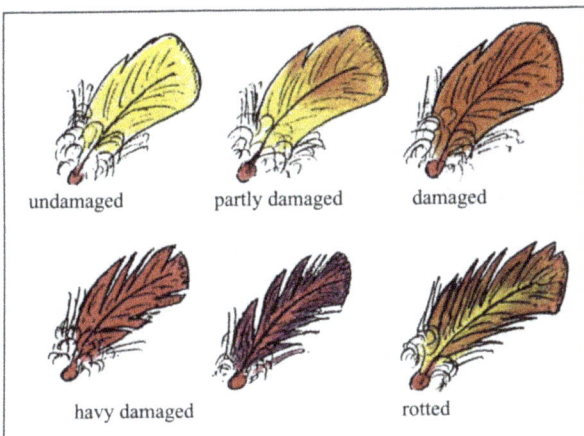

undamaged   partly damaged   damaged

havy damaged   rotted

Fig. 335. Schematic drawing of feathers treated with Neo-carmin W. Undamaged feathers are stained yellow and the quill ends reddish-brown. Damaged feathers, depending on the degree and type of damage, are stained orange, reddish-brown, violet or olive green. (Legend see Fig. 332).

References

1. Bigler N (1965) Praktische Textilmikroskopie – Eine Methode zum Erkennen von Fehlern in Textilien; Textilveredlung 20, No. 9, pp. 559–569

2. Stratmann M (1973) Erkennen und Identifizieren der Faserstoffe, Offprint T 16 from "Handbuch für Textilingenieure und Textilpraktiker", Spohr, Stuttgart

3. Herzog A (1927) Abdrücke tierischer Wollen und Haare in Harz, Melliand Textilberichte 8, p. 341

4. Reumuth H (1955) Gewebeoberflächenstudien. Ein Beitrag zur zerstörungsfreien Textilprüfung und deren Anwendung, Zeitschrift für die gesamte Textilindustrie 57, No. 12, pp. 763–768

5. Bigler N (1960) Die mikroskopische Untersuchung von Schadenfällen an Geweben und Gewirken mit Hilfe des Auflichtes, Durchlichtes und der Oberflächenabdruckmethode mit Gelantineplatten, SVF-Fachorgan 15, No. 4, pp. 251–259

6. Nettelnstroth K (1966) Verfahren zur Präparation von Oberflächenabdrücken, Textilpraxis 21, No. 1, pp. 24–26

7. Jörg F, Neukirchner A (1972) Ein Beitrag über mikroskopische und kinematographische Methoden für die Textilprüfung, Taschenbuch für die Textilindustrie 1972, Schiele & Schön, Berlin, pp. 395–432

8. Mahall K (1980) Charakterisierung von Färbe-, Garn- und Gewebefehlern durch Abdruckverfahren, Textilveredlung 15, No. 1, pp. 373–380

9. Mahall K (1987) Die mikroskopische Untersuchung von Oberflächen textiler Flächengebilde bei Schäden und Auflagerungen, Taschenbuch für die Textilindustrie 1987, Schiele & Schön, Berlin, pp. 442–462

10. Mahall K (1986) Untersuchung von Garnoberflächen im Abdruck, Textilveredlung 21, No. 10, pp. 342–348

11. Nopitsch M (1953) Beitrag zum Nachweis von Schimmel auf Baumwolle und Wollschädigungen im allgemeinen, Melliand Textilberichte 14, pp. 139–142

12. Pauly H, Binz A (1914) Über Seide und Wolle als Farbstoffbildner, 2. Teil, Farben- und Textilindustrie 3, pp. 373–374

13. Agster A (1967) Färberei- und Textilchemische Untersuchungen, Springer-Verlag: Berlin Heidelberg New York, pp. 383

14. Rath H (1952) Lehrbuch der Textilchemie, Springer-Verlag: Berlin Heidelberg New York, p. 184

15. Mahall K (1967) Verarbeitungs- und Ausrüstungsfehler unter dem Mikroskop, SVF-Fachorgan 20, No. 9, pp. 570–587

16. Krais P, Viertel O (1933) Untersuchungen über die Veränderung des Wollhaares während seiner Verarbeitung bis zum fertigen Streichgarntuch, Forschungshefte 14 und 15, Deutsches Forschungsinstitut Textilindustrie, Dresden

17. Mahall K (1985) Beispiele von Schadensfällen aus der Praxis, Deutscher Färberkalender 89, Deutscher Fachverlag, Frankfurt/Main, pp. 184–204

18. Pehl F (1984) Fehler in Textilien und die Ermittlung ihrer Ursachen – Schadensfälle aus der Textilindustrie, Taschenbuch für die Textilindustrie, Fachverlag Schiele & Schön, Berlin, pp. 415–429

19. Mahall K, Goebel I (1988) Fortschritte auf dem Gebiete des Entbastens der Seide, Textilveredlung 23, No. 1, pp. 8–16

20. Mahall K (October 1985) Silk and its treatment, Textile Asia, pp. 95–101

21. Koch P A (1972) Mikroskopie der Faserstoffe, Teildruck T 13 aus "Handbuch für Textilingenieure und Textilpraktiker", Spohr, Stuttgart

22. Mahall K (1986) Katalytische Bleichschäden bei der Veredlung der Baumwolle, Deutscher Färberkalender 90, Deutscher Fachverlag, Frankfurt/Main, pp. 52–56

23. Kirner U (1970) Die modernen Systeme der Peroxidbleiche, Melliand Textilberichte 51, No. 9, pp. 1069–1074

24./25. Agster A (1967) Färberei- und textilchemische Untersuchungen, Springer locations: Berlin Heidelberg New York, p. 321, 359

26. Koch P A, Hefti H (1956) Die Quellung von Baumwollfasern als mikroskopische Schlüsselreaktion zum Nachweis chemischer Schäden, Textil Rdsch. (St. Gallen) 11, p. 512, 645

27. Rüttiger W, Kirner U (1970) Die Geschwindigkeit des Wasserstoffperoxid-Eigenzerfalls als wichtigste verfahrenstechnische Kenngröße in der Textilveredlung, Melliand Textilberichte 51, pp. 1075–1084

28. Mahall K (1983) Fehlererkennung in Textilien mit Hilfe von Abdruckverfahren, Deutscher Färberkalender 87, Franz Eder, Frankfurt/Main, pp. 213–232

29. Mahall K (1974) Schadenfälle an Textilien aus der Praxis, Textilveredlung 9, pp. 43–53

30. Stratmann M (1988) Methoden der qualitativen Faseranalyse (III): Identifizierung der Polyester und Polyolefine, Taschenbuch für die Textilindustrie 1988, Schiele & Schön, Berlin, pp. 450–464

31. Mahall K (1976) Mechanische Schäden an Textilfasern mikroskopisch gesehen, Deutscher Färberkalender 80, Franz Eder, Frankfurt/Main, pp. 378–404

32. Mahall K (October 1985) Silk and its treatment, Textile Asia, pp. 95–101

33. Mahall K (1970) Mikroskopische Untersuchung textiler Reklamations- und Scha-
     densfälle, Textilpraxis 25, No. 10, pp. 608–613

34. Rapp A, Mahall K (1961) Fehler in Textilien, mikroskopisch gesehen, Textilpraxis 16,
     No. 6, pp. 600–607

35. Mahall K (1981) Mikroskopische Befunde zur Faltenmarkierung in der kontinuier-
     lichen Vorbehandlung schwerer Baumwollqualitäten, textil praxis international 36,
     No. 11 and No. 12, pp. 1253, 1254, 1327–1330

36. Mahall K (1969) Hitzeschädigungen an Synthetiks – mikroskopisch gesehen, Textil-
     praxis 24, No. 3, pp. 175–180

37. Mahall K (1988) Die Identifizierung der thermischen Deformierungen und Schäden
     der Synthetiks im Mikrobild, Taschenbuch für die Textilindustrie, Schiele & Schön,
     Berlin, pp. 443–449

38. Mahall K (1984) Strukturell bedingte Streifenbildung in Polwaren – Beispiele aus der
     Praxis, Deutscher Färberkalender 88, Franz Eder, Frankfurt/Main, pp. 164–180

39. Mahall K (1985) Sengfehler – mikroskopisch gesehen, Taschenbuch für die Textilin-
     dustrie, Schiele & Schön, Berlin, pp. 331–334

40. DIN 54208 (December 1975) Quantitative Bestimmung der Anteile binärer Mischun-
     gen; regenerierte Zellulosefasern mit anderen Fasern, besonders Baumwolle, Amei-
     sensäure/Zinkchloridverfahren

41. Schneider R (1959) Kettspritzer und Klemmschüsse in Geweben aus synthetischen
     Fasern, Reyon, Zellwolle 11, pp. 735–738

42. Mahall K (1982) Beispiele aus der praktischen Textilmikroskopie, Taschenbuch für
     die Textilindustrie, Schiele & Schön, Berlin, pp. 324–365

43. Mahall K (1973) Die Streifenbildung in Web- und Wirkwaren, Textilbetrieb 91,
     No. 11, pp. 44–49

44. Mahall K (1984) Garndifferenzen als Ursache von Farbstreifen in textilen
     Flächengebilden, Taschenbuch für die Textilindustrie, Schiele & Schön. Berlin,
     pp. 430–446

45. Mahall K (1975) Aufklärung von Fehlern an Geweben und Wirkwaren unter der Lupe
     bzw. dem Lupenmikroskop, Textilveredlung 10, No. 4, pp. 150–159

46. Mahall K (1971) Farbstoffe als Indikatoren für Fehler in Textilien, Deutscher Färber-
     kalender 75, Franz Eder, Stuttgart, pp. 577–587

47. Reumuth H (1955) ROX-UG, ein Beitrag zum Abdruckverfahren zwecks zerstö-
     rungsfreier Oberflächenprüfung, Melliand Textilberichte 36, No. 6, pp. 533–539

48. Mahall K (1983) Ursachen für die Bildung von Spannfäden in textilen Flächengebilden, Taschenbuch für die Textilindustrie, Schiele & Schön, Berlin, pp. 411–422

49. Mahall K (1984) Nachweis von Auflagerungen auf Textilien mit Hilfe von Abdruckverfahren, Textilbetrieb 102, pp. 36–43

50. Mahall K (1964) Fortschritte auf dem Gebiet der permanent antistatischen Ausrüstung von Hochbauschgarnen aus Polyacrylnitrilfasern für Wirk- und Strickwaren, Wirkerei- und Strickerei-Technik, No. 11, pp. 549–555

51. Mahall K (1982) Biologische Schäden an Textilfasern – mikroskopisch gesehen – Beispiele aus der Praxis, Internationales Textil-Bulletin, No. 4, pp. 280b, 285–286, 291–292

52. Mahall K (1989) Mikroskopischer Nachweis biologischer Schäden, Taschenbuch für die Textilindustrie, Schiele & Schön, Berlin, pp. 195–209

53. Bartsch I (1931) Über den enzymatischen Abbau tierischer Fasern durch Bakterien, Melliand Textilberichte 12, pp. 760–762 und 13 (1932), pp. 21–24

54. Bigler N (1975) Einige mikrochemische Methoden zum Nachweis von Fehlern in Textilien, Textilveredlung 10, pp. 134–150

55. Mahall K (1975) Das Problem der Fleckenbildung in der Färberei und Ausrüstung, Deutscher Färberkalender 79, Franz Eder, Frankfurt/Main, pp. 324–347

56. Nopitsch M (1951) Textile Untersuchungen, Konradin-Verlag Robert Kohlhammer, Stuttgart or Leinfelden-Echterdingen

57. Schulze B, Sommer H (1943) Zur Prüfung der Bakterienfestigkeit von Wolle, Melliand Textilberichte 24, pp. 105–106

58. Horio M and Kondo T (1953) Crimping of Wool Fibres, Text. Res. I 23, pp. 373–386

59. Mahall K (1973) Hochveredelte Geflügelfedern als Füllmaterial für voll waschbare Kissen und Campingartikel, Deutscher Färbekalender 77, Franz Eder, Stuttgart, pp. 253–270

60. Mahall K (1982) Neue Untersuchungen für Geflügelfedern in Hinblick auf die Verwendung in Bekleidungstextilien, Deutscher Färbekalender 86, Franz Eder, Frankfurt/Main, pp. 193–208

Figures

Subject Index

Appendix

Technical Equipment, Chemicals, Reagents and Dyes for Microscopic Damage Analysis

1. **Optical devices** (e.g. Zeiss, Leitz)
   Magnifying glass
   Low-magnification microscope and/or stereo microscope
   Light microscope with polarization equipment and camera
   Variable temperature microscope

2. **Microtome** (Leitz, Jung)

3. **Streak analyzer** for the preparation of large film imprints (Atlas SFTS)
   Two polished steel plates (stainless steel) with the size of microscope slides (7.5 × 2.5, thickness approx. 1 mm) for the preparation of negative imprints in thermoplastic films
   Two plywood plates with the size of microscope slides (7.5 × 2.5, thickness approx. 5 mm)
   Screw clamp
   Drying cabinet
   Polystyrene film, 100 μm (clean, constant thickness)
   Polypropylene film, 30 – 35 μm (clean, constant thickness))

4. **Chemicals**
   Formic acid, conc.
   Ammonia, conc.
   Caustic potash
   Ammonium thiocyanate
   Benzoyl peroxide, 50 %
   Boric acid
   Cover glass mounting cement (Mixture of colophonium and wax, Merck)
   Dioxane
   Acetic acid, 60 %
   Ethanol
   Glauber's salt

Glycerol
Iodine
Potassium hexacyanoferrate (II)
Potassium iodide
Potassium sodium tartrate
Copper sulfate, cryst.
Roskydal E 81 and K 70 (Bayer)
Methanol
Lactic acid
m-cresol
Sodium acetate
Sodium nitrite
Sodium hydroxide, 15%
Pattex Stabilit express (Henkel)
Phenol
Hydrochloric acid, conc, p.a.
Hydrochloric acid, dil., p.a.
Sodium carbonate solution, 10%
Sulfanilic acid p.a.
Zinc chloride, free from water

5. Reagents

Ammoniacal potassium hydroxide: Dissolve 2 g caustic potash in 50 ml conc. ammonia, while carefully stirring and cooling

Cotton Blue-lactophenol
Solution I: 20 ml lactic acid
               20 g phenol
               40 ml glycerol
               20 ml dist. water

Solution II: Dissolve 2 g Lanaperl Blue RN 150 (Hoechst) in 100 ml dist. water.
             The reagent is prepared with 50 ml solution I and 10 ml solution II.

Zinc chloride-iodine (Formulation acc. to Merck, Darmstadt)   Dissolve 66 g zinc chloride, free from water and 6 g potassium iodide in 34 ml dist. water, then add as much iodine as the solution absorbs.

Fehling's solution	(a) Dissolve 34.5 g cryst. copper sulfate in 500 ml dist water. (b) Dissolve 173 g potassium sodium tartrate and 70 g caustic soda in 500 ml dist. water. For application, equal quantities of (a) and (b) are mixed and diluted 1:3 with dist. water.
Iodine/Dioxane/ Boric acid	20 ml n/10 iodine solution, 80 ml dioxane, 1 g boric acid. Dissolve boric acid while heating, cool the solution down to 25 °C and add dioxane to obtain 100 ml. Finally, add 7 ml dist. water.
Iodine solution	Dissolve 1 g iodine and 3 g potassium iodide in 500 ml dist. water.
m-cresol/Fat Red	0.5 g Fat Red 5B (Hoechst) are wetted with a small quantity of methanol and dissolved in 50 ml m-cresol.
Cuoxam	Electrolyte copper dust is mixed with conc. ammonia. Decant from the deposit as soon as the solution changes to dark blue. Its shelf life is limited.
Pauly reagent	Sulfanilic acid is diazotized. The resulting diazobenzene sulphonic acid is collected on a filter and dissolved by pouring 10% ice-cold sodium carbonate solution over it, see chapter 2.1.1.
Sudan Red/Glycerol/ Alcohol	Dissolve 0.3 g Sudan Red 460 in 50 ml ethanol and add 50 ml glycerol.

6. Dyes (former and new name/supplier)

tests (page...)	former name/supplier	new name/new supplier/ substitute
micro organism (p. 196)	Lanaperl Blue RN 150/ Hoechst	Telon Blue AGLF/ DyStar
red/green test of mature cotton (p. 48)	Diphenyl Red 5B 182%/Ciba-Geigy	Levacell Red 4B powder/ DyStar Pergasol Red 2B 182%/ Ciba SC
	Solophenyl Green BL/ Ciba-Geigy	Cibafix Green E-B/ Ciba SC
Nonax 1166 on textile fibres (p. 176)	Sirius Pink BB 143%/ Bayer	Sirius Red Violet RL/ DyStar
Fat/Oil/Wax		Duranol Blue PP/ICI
Detection of saponified acetate fibers (p. 66+67)	Sirius red 4B/Bayer	Levacell red 4B/Bayer (paper dye)
Staining of gum (p. 42)	Sirius Red F3 B 200%/ Bayer	Sirius Red F3 B/ DyStar
Wool anti-felt and un-treated (p. 37)	Supramin Yellow GW/ Bayer Acilan Fast Navy Blue R/ Bayer	Telon Yellow GW/ DyStar (acid yellow 61) Sandolan navy P-R2/ Sandoz (acid blue 92)

Unchanged test colours :

Methylen Blue (CI 45170)/Merck, Darmstadt
Rhodamine B (CI 52015)/Merck, Darmstadt
Neocarmin W/Fesago, Chem. Fabrik Dr. Gossler, 69207 Sandhausen, Germany
Oxycarmin/Fesago, Chem. Fabrik Dr. Gossler, 69207 Sandhausen, Germany